Common Core Biology

The Basis for Life

Monica Sevilla

Contents

What are Living Things?

What is a Cell?
Types of Cells
What is a Plant Cell?

Parts of the Cell

The Basic Molecules of Life
The Essential Elements of Life
The Early Earth and the Making of the Molecules of Life
What is Photosynthesis?

Light
Cellular Respiration

The Basic Molecules of Life

What is DNA?
What is RNA?
What are proteins?

What is an amino Acid?

What is a gene?
Gene Regulation
What is a Genetic Mutation?

Transcription

Translation

What is a Cell?
Types of Cells
The Plant Cell
Parts of the Cell
The Cell Membrane
The Nucleus of the Cell
The Mitochondrion
The Endoplasmic Reticulum Golgi Apparatus

The Cell Organelles and their Functions Study Cards
The Structure of Animal and Plant Cells Activity

What is Photosynthesis?

An Introduction The Chloroplast and the Process of Photosynthesis
Cellular Respiration

What is ATP
Transcription
Translation
Mitosis and the Cell Cycle
Cell Division (The Cell Cycle) Study Cards Meiosis: Producing Gametes

Cell Differentiation

The Cell Unit Exam

What is Evolution?
What is an Adaptation?

What is Natural Selection?
Adaptive Radiation
Speciation
The Evolution of Fish
The Evolution of Amphibians
The Evolution of Reptiles
The Evolution of Mammals
The Evolution of Birds
The Evolution of Primates
The Transition from the Jungle to the Savanna

What is Evolution?
Primate Evolution
What are Adaptations?
The Transition from the Jungle to the Savanna T

The Transitional Species:Homo Habilis and Homo Rudolfensis

Homo Erectus
Homo Heidelbergensis
Homo Neanderthalensis
Differences in Adaptations among Neanderthals and Homo Sapiens
The Denisovans
Homo Floresiensis
Coexistence: Homo Sapiens and Other Hominids

Common Core Biology

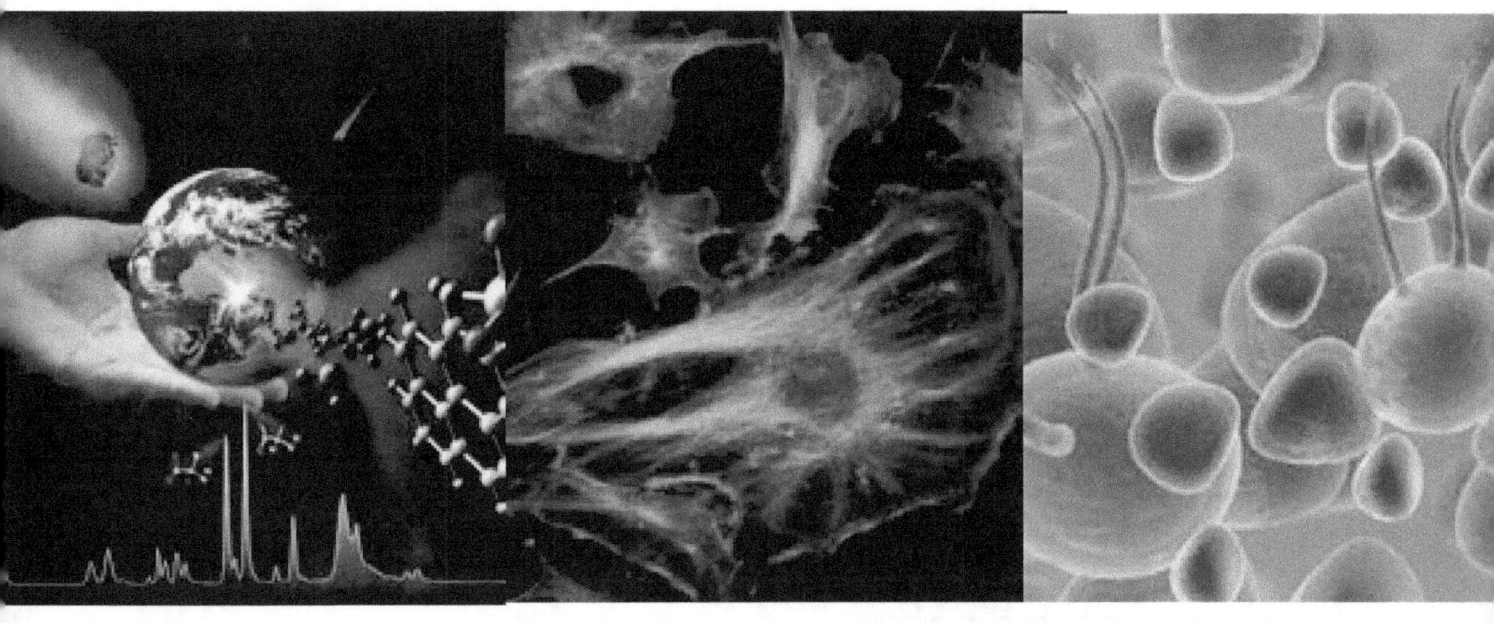

Cells
and the Molecules
of Life

Monica Sevilla

Contents

What are Living Things?
What is a Cell?
Types of Cells
What is a Plant Cell?
Parts of the Cell
The Basic Molecules of Life
The Essential Elements of Life
The Early Earth and the Making of the Molecules of Life
What is Photosynthesis?
Cellular Respiration

What are Living Things?

When we study **biology**, we are actually studying about **organisms** or living things. Organisms include plants, animals, and microbes. These organisms are **classified** or grouped into 5 **kingdoms** of living things: plantae, animalia, fungi, monera, and protista.

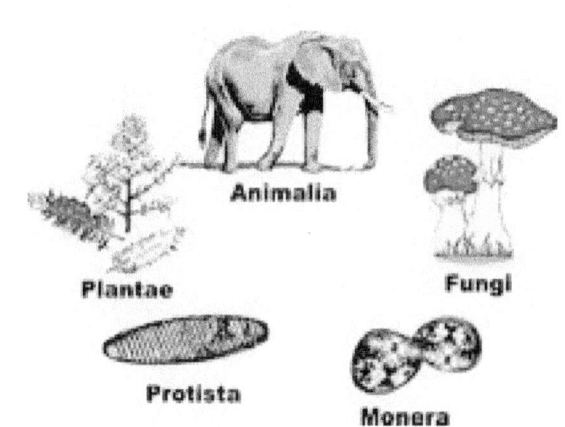

It is believed by scientists that all organisms are the decendents and evolved from one living creature that existed billions of years ago. Because of this continuous evolution and the changes that were made within the DNA during this time span, all living organisms have some certain characteristics in common.

They have cells.
They sense and respond to change.
They can reproduce.
They have DNA.
They use energy.
They grow and develop.

Most organisms have the same **basic needs**. These needs include: water, air, a place to live, and food.

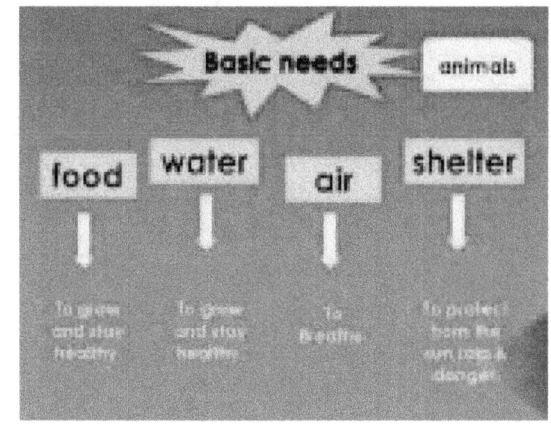

If these needs are not met, organisms may get ill or even die. The survival of all organisms depend on these four basic needs to ensure that they reach adulthood and their able to successfully reproduce in order to keep their species from going extinct.

Knowledge and Comprehension
Words to Know:

Biology:

Classified:

Organisms:

Kingdoms:

Basic Needs:

1. What is an organism?

2. What are the five kingdoms organisms are classified into?

Application, Analysis, Evaluation and Synthesis

3. What are the common characteristics of all organisms.

4. Explain why organisms today have these common characteristics.

5. What are the basic needs of living things? Why is it important to meet these needs?

What is a Cell?

A **cell** is the smallest functional and structural building block of life. Cells make up all living things. They communicate with each other and work together to make up the structure of living things and to carry out the many functions of organism.

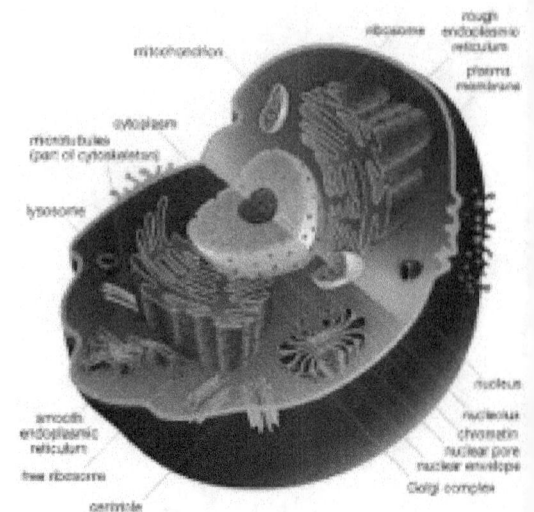

The first person to describe the cell from his observations was **Robert Hooke** in the 1660's. In 1665, he built a microscope to look at objects that could not be seen with the naked eye. A **microscope** is a scientific tool that use glass lenses to magnify tiny objects. The first observation he made was looking at a sample of dead and dry bark from a cork tree. What he saw were many little rooms or "cells" that were held closely together.

The next observation Hooke made was looking at a slice of a living plant. He also saw cells in this sample as well. He noticed that these cells had fluid inside them and called them "juicy." He also notice that plants cells have a **cell wall** or the outermost structure of

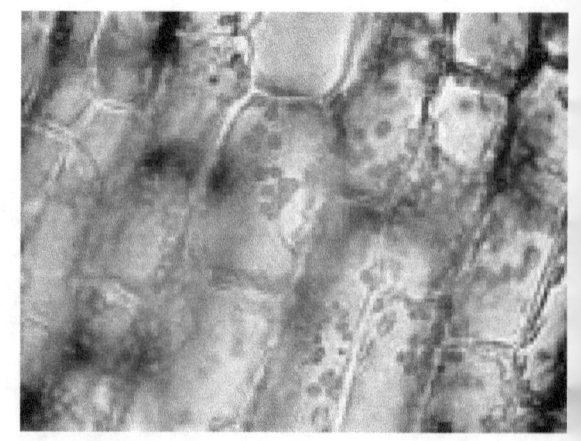

the cell.

Anton van Leeuwenhook, in 1673, built and used his own microscope to look at a sample of scum that he found in a pond. He noticed that there were small organisms swimming in the water. He called these organisms "little animals" or animalcules. We now know these organisms as **protists** or single celled animals. He was also the first person to use his microscope to see bacteria, a type of single celled animal.

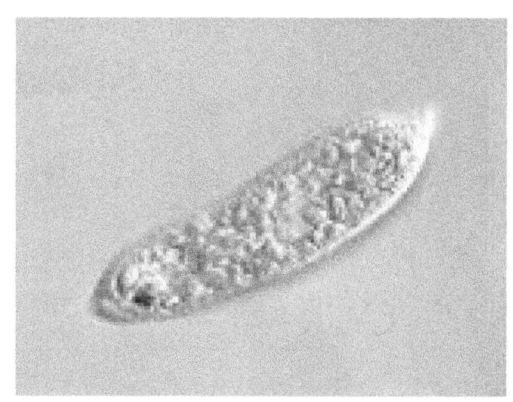

**Knowledge and Comprehension
Words to Know:**

Cell:

Robert Hooke:

Cell Wall:

Anton van Leeuwenhook:

Protists:

Microscope:

1. What is a microscope?

2. What can you see with a microscope that you can not see with your unaided eye?

Application, Analysis, Evaluation and Synthesis

3. Describe the contribution that Robert Hooke made to science. What did he discover and why was it important.

4. Describe the contribution that Anton van Leeuwenhook made to science. What did he discover and why was it important.

5. Compare and contrast the contributions that were made by Robert Hooke and Anton van Leewenhook. What was similar about their discoveries? What was different?

6. If the discovery of the microscope had not been made, how would this have impacted or affected our world?

Types of Cells

A **cell** is the smallest functional and structural building block of life. Cells make up all living things. They communicate with each other and work together to make up the structure of living things and to carry out the many functions of organism.

There are two different types of cells. Cells that have a nucleus and cells that do not have a nucleus. The **nucleus** is a structure within the cell that stores all the genetic information known as **DNA** for the function and reproduction of the cell.

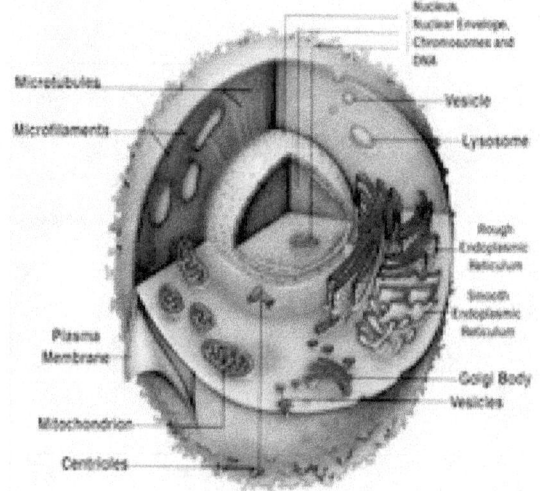

Eukaryotic cells have a nucleus. They have genetic material DNA. Plant cells and animal cells are examples of eukaryotic cells.

Prokaryotic cells or prokaryote do not have a nucleus. Prokaryotic cells are classified into two groups: bacteria and archaea.

It is believed by scientists that the

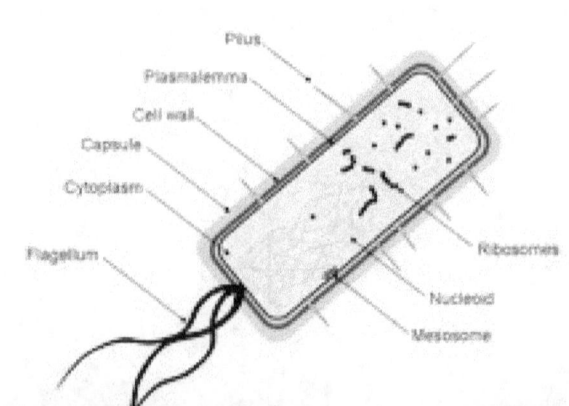

first living things on Earth were probably prokaryotes. More complex life forms are said to have evolved from these single celled organisms. The evidence supports this fact is that microbials, 3.48 billion years old, were discovered in western Australia. Other evidence that supports this is the research that has been done by biochemist Douglas Theobaldis who calculated that the last universal common ancestor of all life forms on Earth is lease 10^{2860} more probable than having multiple ancestors. This evidence suggests that life on Earth followed an evolutionary pattern suggested by Charles Darwin in his book **On the Origin of Species**. In his book, he infers that all living organisms on Earth arose through evolutionary processes from one primordial form.

Sources:

http://news.nationalgeographic.com/news/2010/05/100513-science-evolution-darwin-single-ancestor/

Knowledge and Comprehension
Words to Know:

Cell:

Nucleus:

DNA:

Eukaryotic Cells :

Prokaryotic cells:

1. What is a cell?

2. What are the two major types of cells?

Application, Analysis, Evaluation and Synthesis

3. Explain how eukaryotes and prokaryotes are different from each other. What do these organisms have in common?

4. Could a living organism survive without a nucleus? Find evidence from the text to support your answer.

5. Do you agree with the claim "All complex organisms evolved from a single-celled organism." Find evidence from the text to support your answer.

What is a Plant Cell?

Plants are **eukaryotes** or a multicellular organisms. **Plant cells** have unique features when compared to animal cells. They have most of the organelles and structures that exist within the animal cell, but also have a cell wall, chloroplasts, and a large central vacuole that have specialized functions to carryout the process of photosynthesis. **Photosynthesis** is the process of a plant making the simple sugar glucose with the light energy from the sun.

Plants have organelles called **chloroplasts** which is the location of photosynthesis. The chloroplasts use carbon dioxide and water molecules and convert them glucose through a series of chemical reactions. The chloroplasts contain **thylakoid membranes** which house the green pigment chlorophyll. **Chlorophyll** absorbs or takes in light energy within its molecular bonds. Electrons are then transferred from chlorophyll molecules to reaction centers

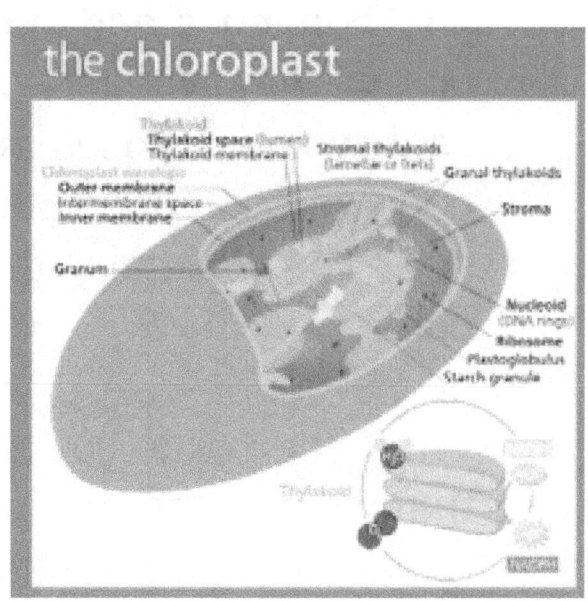

where it is used in photosynthesis.

The **cell wall** is a structure that surrounds the cell membrane of plant cells. The cell wall is rigid. It maintains the shape of the cell and protects the cell fro damage. Cell walls allow fluids and molecules to pass in and out of the cell. Unlike the animal cell, which is made of phospholipids and proteins, the cell wall is made mostly of a carbohydrate called **cellulose** and protein.

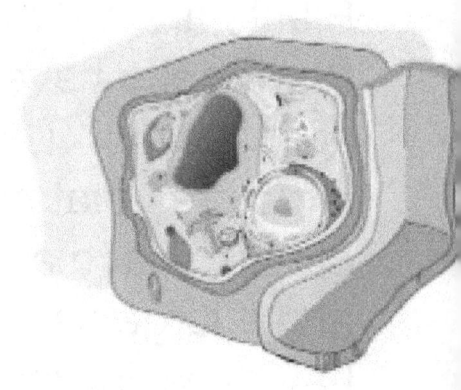

The **large central vacuole** maintains the internal pressure of fluids within the plant cell. This structure is a water-filled membrane that regulates the amount of fluid coming in and out of the cell. The large central vacuole draws water into itself when it senses that the concentration of solutes is low outside cell. This maintains the rigidity of the cell.

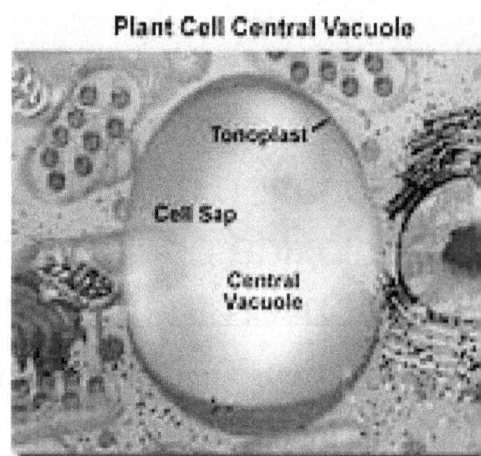

When the solute concentration is higher on the inside of the cell then the outside, the large central vacuole will release water out of the cell in order to decrease the concentration. In the plant cell, a solute is a mixture of

water and other substances such as molecules and minerals. These two actions are how the cells respond to changes in their environment in order to maintain equilibrium. **Equilibrium** is a state where the environments both inside and outside of the cells are in balance. In this case, the cells try to balance the solute concentration, and the pH of the solution through the movement of water in and out of the cells through a semi-permeable membrane. This process is known as **osmosis**.

Knowledge and Comprehension
Words to Know:

Eukaryotes:

Photosynthesis:

Chloroplasts:

Chlorophyll:

Thylakoid membranes:

Cell Wall:

Large Central Vacuole:

Cellulose:

Equilibrium:

Osmosis.

1. What three organelles or structures are unique to the plant cell when compared to an animal cell?

2. What is osmosis?

Application, Analysis, Evaluation and Synthesis

3. What function does the plant cell serve? How are the cell wall in a plant and the outer membrane of the animal cell different?

4. How does the plant cell maintain equilibrium?

5. If the solute concentration is higher in side the cell than it is outside the cell, what must occur for the cell to achieve equilibrium?

6. Predict what would occur if the central vacuole inside the plant cell was damaged? What would happen to the cell?

7. What is a chloroplast and what is its function within the plant cell? Could the cell function without the chloroplast? why or why not?

8. What is chlorophyll and what is its function in the plant cell? What is its role in the process of photosynthesis.

Parts of the Cell

A **cell** is the smallest functional and structural building block of life. They communicate with each other and work together to make up the structure of living things and to carry out the many functions of organism. Animal cells are eukaryotic cells. **Eukaryotic cells**, unlike prokaryotic cells, have a nucleus. They also have a host of organelles that carry out the functions and processes that occur within the cell.

Nucleus: is a large organelle inside the cell where genetic material or DNA is stored and replicated (copied). A dark spot within the nucleus, called the nucleolus, is the location where ribosomes, organelles that make proteins, are made.

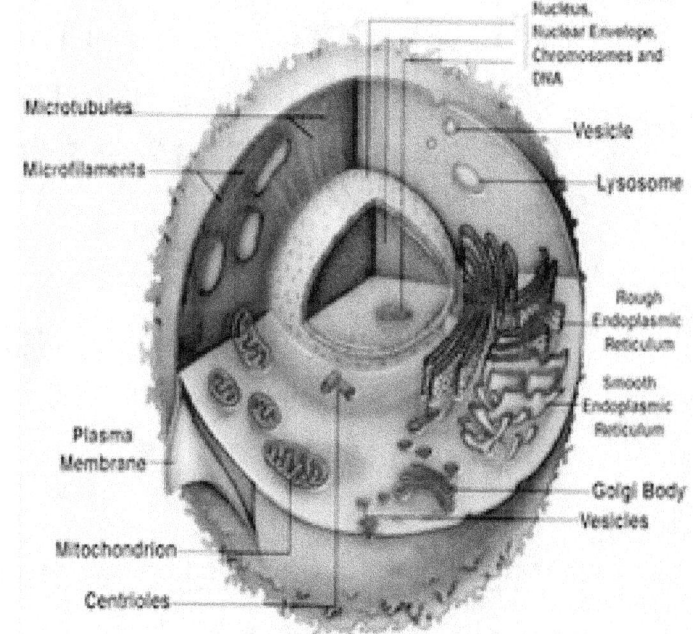

Cell Membrane: the cell membrane is the outermost layer of the cell. It is made up of phospholipids (fats) and proteins. It protects the cell and regulates the materials that come in and out of the cell.

Cytoskeleton: is a web of proteins in the cytoplasm of the cell that provides a framework and structure for the organelles within the cells. It also determines the shape of the cell.

Ribosomes: organelles that make proteins for the cell.

Endoplasmic Reticulum or ER: is a system of folded membranes in which proteins, lipids, and other materials. The endoplasmic recticulum transports and delivers substances to different places in the cell.

Mitochondria: is the powerhouse of the cell. It is an organelle that breaks down sugar to release energy. This energy is stored in the adenosine triphosphate (ATP) molecule.

Golgi Complex: is an organelle that packages and moves proteins to where they are needed both within and out of the cell.

Vesicle: is a small sac surrounds substances and moves them in and out of the cell.

Lysosome: are vesicles responsible for digestion that found in animal cells. They contain specialized proteins called enzymes that get rid of waste products, foreign invaders, and organelles that have been worn out or damaged.

Knowledge and Comprehension
Words to Know:

Cell:

Nucleus:

Ribosomes:

Endoplasmic Reticulum:

Mitochondria:

Chloroplasts:

Golgi complex:

Vesicle:

Lysosomes:

Eukaryotic Cell:

1. What is a eukaryotic cell?

2. What is the purpose of organelles within the cell?

3. Identify each of the organelles by the following descriptions:

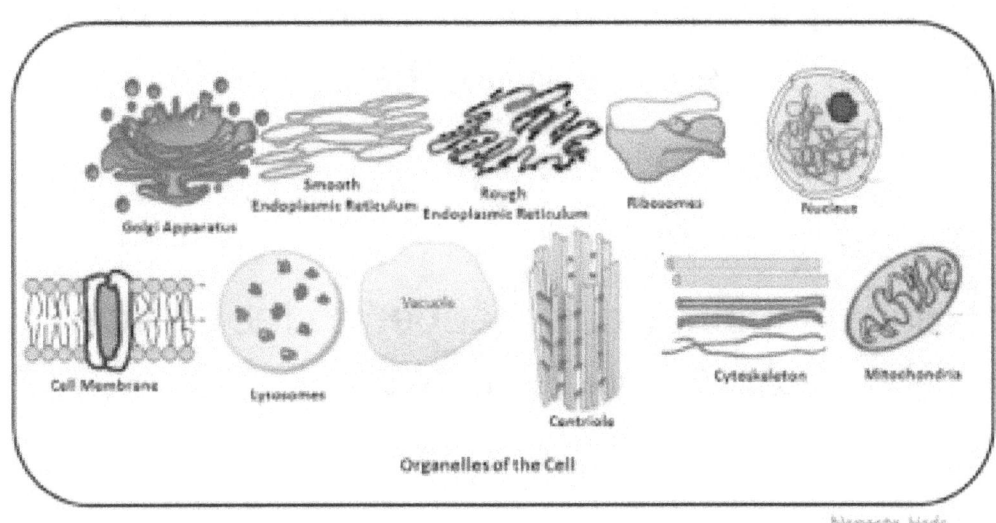

Organelles of the Cell

a. In which organelle is the genetic material if the cell contained within?

b. Which organelle digests waste and worn down organelles?

c. Which organelle processes and transports substances in and out of the cell.

d. Which organelle makes the energy for the cell?

e. Which organelle makes proteins inside it?

f. Which organelle that makes lipids, proteins, and packages proteins for the golgi complex.

g. What cell structure provides the framework for the organelles inside the cell?

h. What cell structure is made of lipids and proteins and regulates what comes in and out the cells?

4. What, do you think, would happen to the proteins within the cell if the nucleus was damaged?

5. The endoplasmic reticulum makes lipids and packages protein for another organelle within the cell. What would happen to the function of this organelle if the endoplasmic reticulum overproduced lipids and proteins?

6. What would happen within the cell if lysosomes did not exist? What would happen to the wastes, foreign invaders, and damaged organelles? Would the cell become toxic? Explain.

7. Why, in your opinion, does the cell have a cell membrane? Why is this important?

The Basic Molecules for Life

The cells within organisms use atoms as building blocks for making molecules. **Atoms** are the basic units of elements that can not be divided or broken down. **Molecules** are atoms that are bonded together. Most molecules within living things are made up of the atoms of six different types of elements: sulfur, nitrogen, potassium, hydrogen, oxygen, and carbon. These atoms combine together in different arrangements to forms macromolecules.

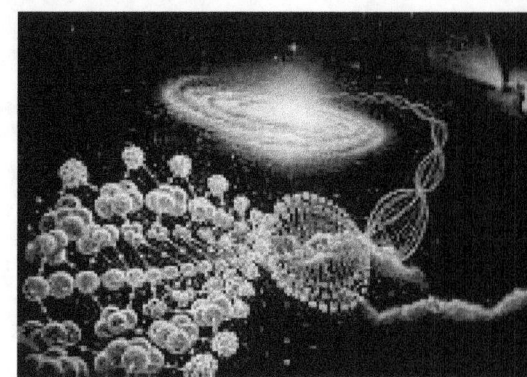

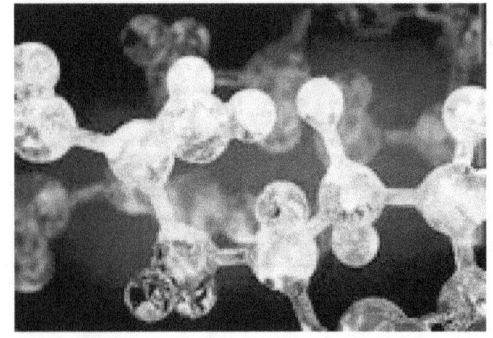

Macromolecules are larger molecules, usually made up of chains of smaller molecules, that the cells either synthesize or make themselves or breakdown and use.

Macromolecules allow the cells to function properly, carry out important chemical reactions, and reproduce. Macromolecules are also classified into four different types: carbohydrates, lipids, nucleic acids, and proteins.

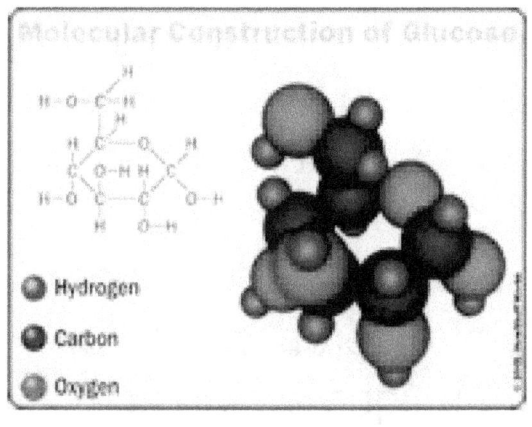

Carbohydrates: are sugar molecules that store energy. When carbohydrates are broken down by the cell, the bonds between the atoms break, and release energy that is used by the cell to fuel different functions.

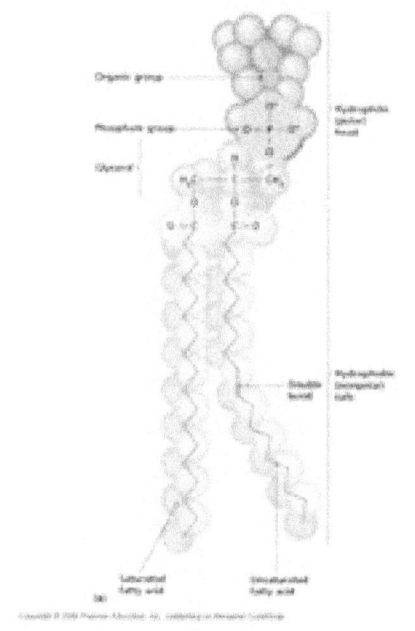

Lipids: are fats that do not dissolve in water. Lipids are classified into different types that include oils, phospholipids, steroids and waxes. Some lipids form barriers against microbes and viruses, prevent loss of water, are used to make hormones, and store large amounts of chemical energy that can be used in the future.

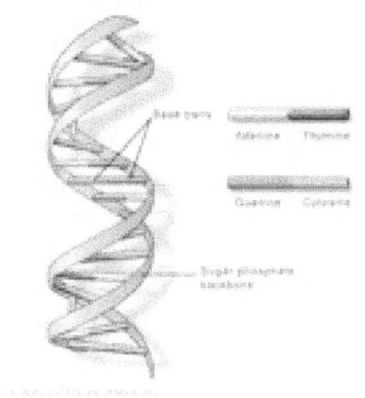

Nucleic Acids: are long chains of molecules called nucleotides. Deoxyribonucleic acid made up of 4 nitrogenous bases: adenine, thymine, guanine, and cytosine. DNA is the genetic material contained within the nucleus of the cell.

**Proteins**: Proteins are long chains of amino acids that have been folded to create a compact structure. They are made from instructions that are coded on the DNA molecule. There are thousands of different proteins, and each has a unique amino acid sequence. They have many different functions and act as catalysts for chemical reactions that occur within organisms.

Knowledge and Comprehension
Words to Know:

Atoms:

Molecules:

Macromolecules:

Carbohydrates:

Lipids:

Nucleic Acids:

Proteins:

1. What is a macromolecule?

2. What four important macromolecules found in organisms.

Application, Analysis, Evaluation and Synthesis

3. What is the difference between a molecule and a macromolecule? What is the relationship between the two?

4. Describe what a nucleic acid is and what a protein is. What is the relationship between nucleic acids and proteins?

5. What do you think would happen to a protein if there was a mistake in the nucleic acid that serves as the genetic instructions for its construction?

6. Identify what the functions lipids are. Why are these functions important?

The Essential Elements for Life

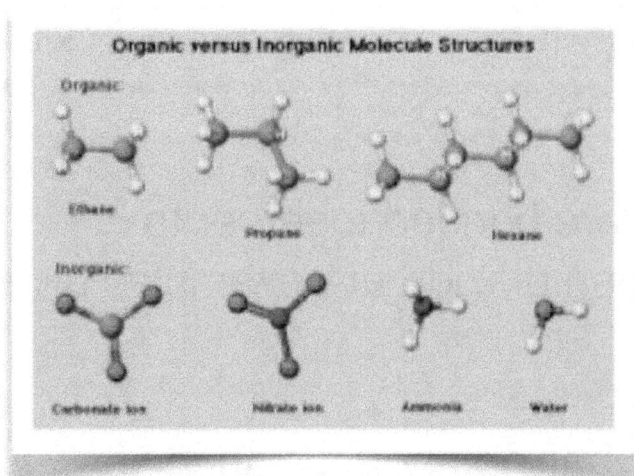

All lifeforms on Earth are made up of organic compounds. **Organic compounds** are compounds composed of carbon atoms either bonded to other carbon atoms to make chains or bonded to different atoms. **Compounds** are atoms that have bonded together to create a substance. Bonds are created through the attraction of positive and negative charges or the sharing of electrons between atoms.

Compounds can either be ionic or covalent. **Ionic compounds** are created from ions or charged atoms, positive and negative atoms, that bond to each other. **Covalent compounds** are created when atoms join together and share electrons with each other.
The most common elements that carbon bonds to include hydrogen, oxygen, and nitrogen. This is because carbon is able to share up to four electrons with other atoms and create **covalent bonds**.

Carbon can make one bond to hydrogen, 2 bonds to oxygen, and 3 bonds to nitrogen. **Inorganic compounds** are substances that do not contain carbon atoms that are bonded together to create chains. Water and carbon dioxide are examples of inorganic molecules.

Carbon can form a variety of small compounds such as carbon dioxide, as well as, **macromolecules**, or massive compounds, such as carbohydrates and lipids. Carbohydrates are molecules that are easily broken down with the body of living organisms to produce energy. Lipids are fats or oils that are an important component of the cell membrane, and used to produce cholesterol and hormones.

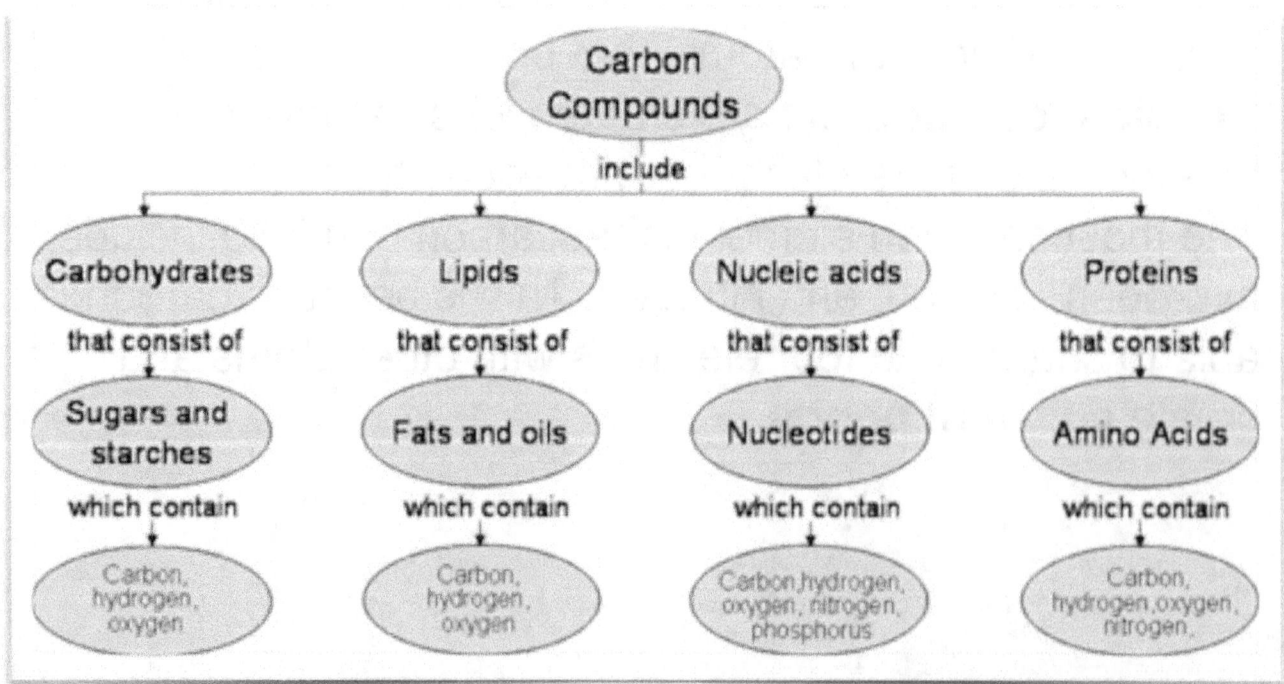

Source:

http://www.sciencelearn.org.nz/Contexts/Just-Elemental/Science-Ideas-and-Concepts/The-essential-elements

Knowledge and Comprehension
Words to Know:

Compounds:

Organic Compounds:

Inorganic Compounds:

Ionic Compounds:

Covalent Compounds:

Macromolecules:

1. What are organic compounds?

2. What is a macromolecule?

Application, Analysis, Evaluation and Synthesis

3. Explain what the major difference is between organic and inorganic compounds.

4. Explain what the major difference is between ionic and covalent compounds.

5. Identify which major elements make up the lifeforms we see on Earth. Which element is the most important? Explain why this element is important.

6. Which macromolecules are the most important to the function and structure of living organisms on Earth? What do these substances have in common? How are they different? Use evidence from the text to support your answer.

The Early Earth and the Making of the Molecules of Life

All forms of life are composed or made up of organic compounds. **Organic compounds** are compounds that are composed of carbon atoms either bonded to other carbon atoms to make chains or bonded to different atoms. **Compounds** are atoms that have bonded together to create a substance. Bonds are created through the attraction of positive and negative charges or the sharing of electrons between atoms. **Inorganic compounds** are substances that do not contain carbon atoms that are bonded together to create chains. Water and carbon dioxide are examples of inorganic molecules.

In 1953, Stanley Miller and Harold Urey, two chemists from the University of Chicago, conducted an innovative experiment to **simulate** or reproduce the environmental conditions that were present during the early formation of the Earth. What they set out to prove was that organic compounds that are necessary for life could have been synthesized or made from simple inorganic molecules such as water, methane gas, ammonia, and hydrogen gas. The chemicals methane, ammonia, and hydrogen were placed in individual flasks and connected in a loop with one flask half filled with water.

The liquid water was heated and evaporated to create a mini-atmosphere that contained water vapor. Two electrodes fired a electrical current, creating lightning, through the water vapor. The water vapor was cooled down to allow it to condense and drip back into the original flask. This process was allowed to occur as an ongoing, continuous cycle. Within the span of one day, a pink mixture was created. 10-15% of the chemicals existed in the form of organic compounds at the end of 2 weeks time. 18% of the methane was converted into organic compounds.

2% of the carbon within the reaction was transformed into amino acids. **Amino acids** are organic compounds that are joined together to create proteins within living cells. The amino acid **glycine** was present in the most concentration. 11 of the 20 amino acids essential for life were created in this experiment. Sugars, such as ribose and acetic acid, a fatty acid, appeared as well.

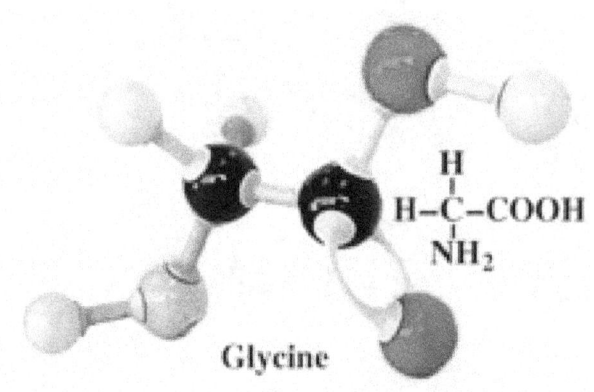
Glycine

This experiment demonstrated that simple inorganic compounds, under the proper environmental conditions, could produce the macromolecules essential for the creation of life on Earth. They were able to prove that the early Earth had the raw materials and the ability to produce organic molecules through specific chemical reactions. They were also able to show that the molecules such as amino acids and sugars such as ribose are important building blocks for life that make up the lifeforms that have existed in the past and exist today. and that they could be made from water and gases with the addition of electrical energy.

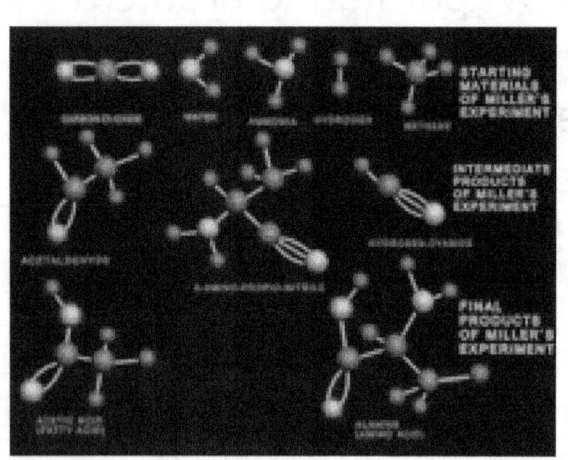

Source:

http://en.wikipedia.org/wiki/Miller%E2%80%93Urey_experiment

Knowledge and Comprehension
Words to Know:

Compounds:

Organic Compounds:

Inorganic Compounds:

Simulate:

Amino Acids:

Glycine:

1. What is the difference between inorganic and organic compounds?

2. Which type of organic molecules were produced during this experiment?

Application, Analysis, Evaluation and Synthesis

3. Which inorganic compounds were present before the experiment began? What was the result of this experiment after 1 week?

4. Which organic compounds were present 2 weeks after the experiment began? What types of macromolecules were among these substances and are important for life to exist?

5. Find evidence from the text to support the claim: Organic chemicals, the molecules important for life, can be made from a few inorganic compounds under the right conditions.

6. Explain what the purpose of this experiment was. Did the scientists achieve their goal? Explain.

**Knowledge and Comprehension
Words to Know:**

-

-

-

:

:

1.

2.

Application, Analysis, Evaluation and Synthesis

What is Photosynthesis?
An Introduction

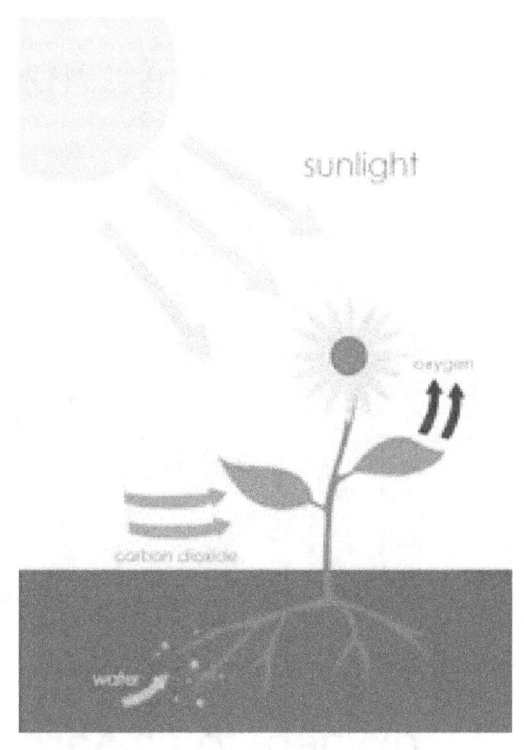

Photosynthesis is a process where photo or light energy is converted or changed by plants, cyanobacteria, and photoautotrophs such algae to **synthesize** or make glucose. An **autotroph** is an organism that makes its own food. A **photoautotroph** is an organism that makes its own food through photosynthesis. **Glucose** is a carbohydrate, a simple sugar, that is made of 6 carbon atoms that are arranged into a ring structure. This glucose is stored within plants and organisms and used as food.

Photosynthesis requires carbon dioxide (CO_2) from the atmosphere and water (H_2O) and light energy as the raw materials for this process. Sugars are produced through chemical reactions that are dependent on light as a source of energy. As the sugars are formed, oxygen gas is produced as a waste product. The following equation illustrates how many molecules of carbon dioxide and water are used to synthesize or make one molecule of glucose.

$$6CO_2 + 12H_2O + light \rightarrow C_6H_{12}O_6 + 6O_2 + 6H_2O$$

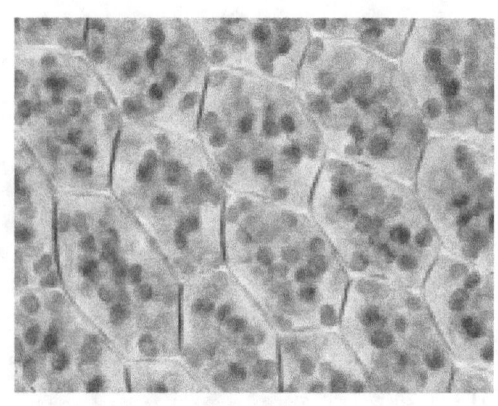
Scientists believe that the atmosphere of the early Earth became oxygen-rich through photosynthetic bacteria and algae living in the oceans millions of years ago. This transformation caused a major shift from a global **anaerobic** environment (no oxygen) to an **aerobic** (oxygen filled) environment, setting the stage for more complex animals and plants to evolve.

Knowledge and Comprehension
Words to Know:

Photosynthesis:

Synthesis:

Autotroph:

Photoautotroph:

Glucose:

Anaerobic:

Aerobic:

1. Describe what an autotroph is.

2. What is the difference between an autotroph and a photoautotroph.

Application, Analysis, Evaluation and Synthesis

3. What is photosynthesis. Summarize the process in your own words.

4. Which organisms are considered photoautotrophs? Do you think the process of photosynthesis is the same in each of these organisms? Why or why not?

5. How do scientists believe that oxygen was first formed on Earth? Do you agree with this theory? Why or why not?

What is Light?

Light is a type of radiant energy that is classified as electromagnetic in nature. **Electromagnetic energy** is made up of 2 types of energy fields: electric fields and magnetic fields. An **electric field** is created by an electric charge and a magnetic field. A **magnetic field** is the magnetic influence of electric currents and materials that are magnetic. This explains why light has properties that are both electric and magnetic in nature. The two energy fields exist together and are perpendicular or at 90 degree angles in relation to each other. Because of this, light behaves like a wave.

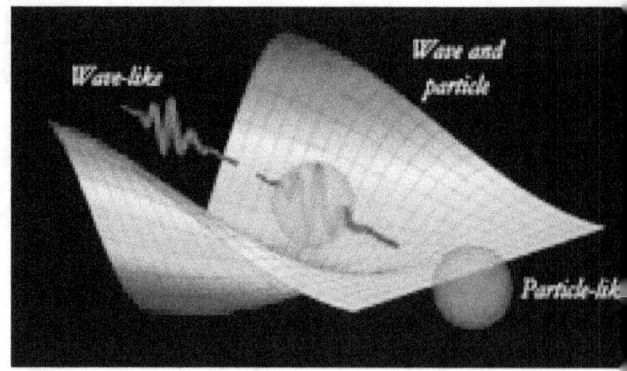

Light energy also behaves as a particle. This is because light is made up of **photons** which are particles of energy that exist and move in **quanta** or small packages. The photon is the force carrier for the electromagnetic force. A force carrier is a particle that causes forces to occur between other particles.

electromagnetic radiation from a star is created by nuclear fusion reactions at high temperatures within the its core. Hydrogen atoms are fused together to form helium atoms. During this reaction positrons and energy, in the form of light photons, are released. This electromagnetic radiation includes visible light, non-visible radiation (such as UV or ultraviolet radiation) radiation and positrons.

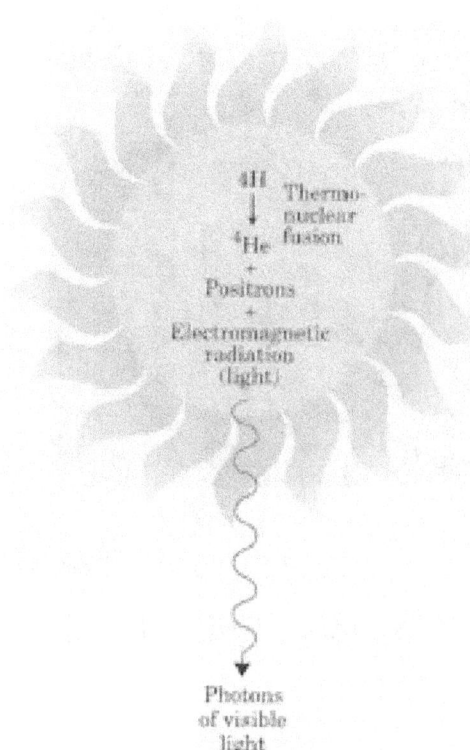

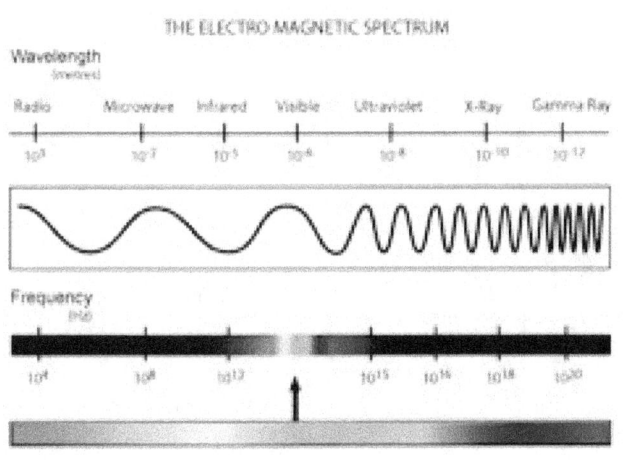

Visible light can be seen because its energy encompasses the **visible light spectrum**. Light is visible to the human eye at the wavelengths of 400-700 nanometers. It is generated by nuclear fusion reactions that occur in stars. Light reaches Earth in 8 minutes traveling at the speed of 3×10^8 meters/second.

 Albert Einstein, through his research, was able to show that the light from stars could be affected by the sun's gravity. He was able to demonstrate that starlight could be bent due to the gravity of a celestial body such as a star, a planet, or a black hole. He theorized that gravity affects the fabric of space-time around a heavenly body by warping it. He also predicted that the rotation of a heavenly body also affects space-time by twisting it.

Focus Questions

1. What is light?

2. What is electromagnetic energy? Is all electromagnetic energy visible?

3. How is a photon created in a star?

4. Describe what makes visible light "visible."

5. Does light act like a wave or a particle? Explain.

6. Can light be affected by gravity? Explain.

Cellular Respiration

Cells, to have enough energy to carry out their functions properly, need to utilize energy on an ongoing basis. To provide this ongoing source of energy, cells use the process of cellular respiration. **Cellular respiration** is a set of chemical reactions that catabolize and convert macromolecules such as Amino acids, fatty acids, and the simple sugar **glucose** into useable energy to fuel the many activities of the cell. The activities that are powered by cellular respiration include the transport of molecules across the cell membranes, locomotion, and the synthesis or making of molecules and macromolecules in the cell.

The goal of cellular respiration is to **catabolize** or break down macromolecules and to release the energy stored within its high energy molecular bonds. During **glycolysis**, glucose molecules are broken down in the presence of oxygen, which acts as an **oxidizing agent** or an electron acceptor. Glucose molecules are converted into the high energy molecules **pyruvate** or CH_3COCOO^- and hydrogen ions (H^+).

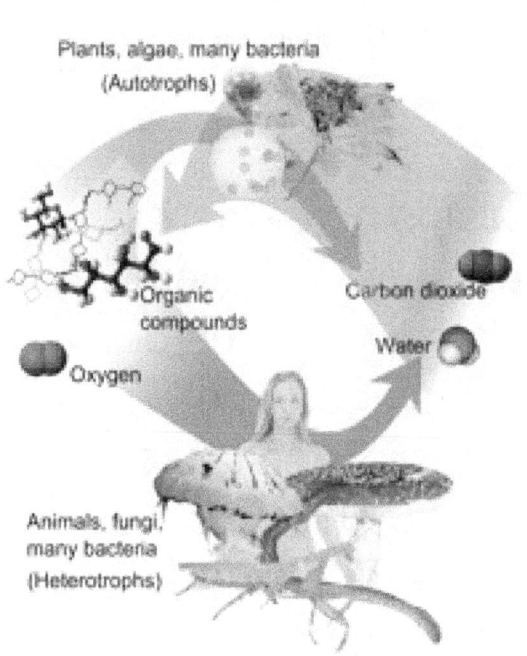

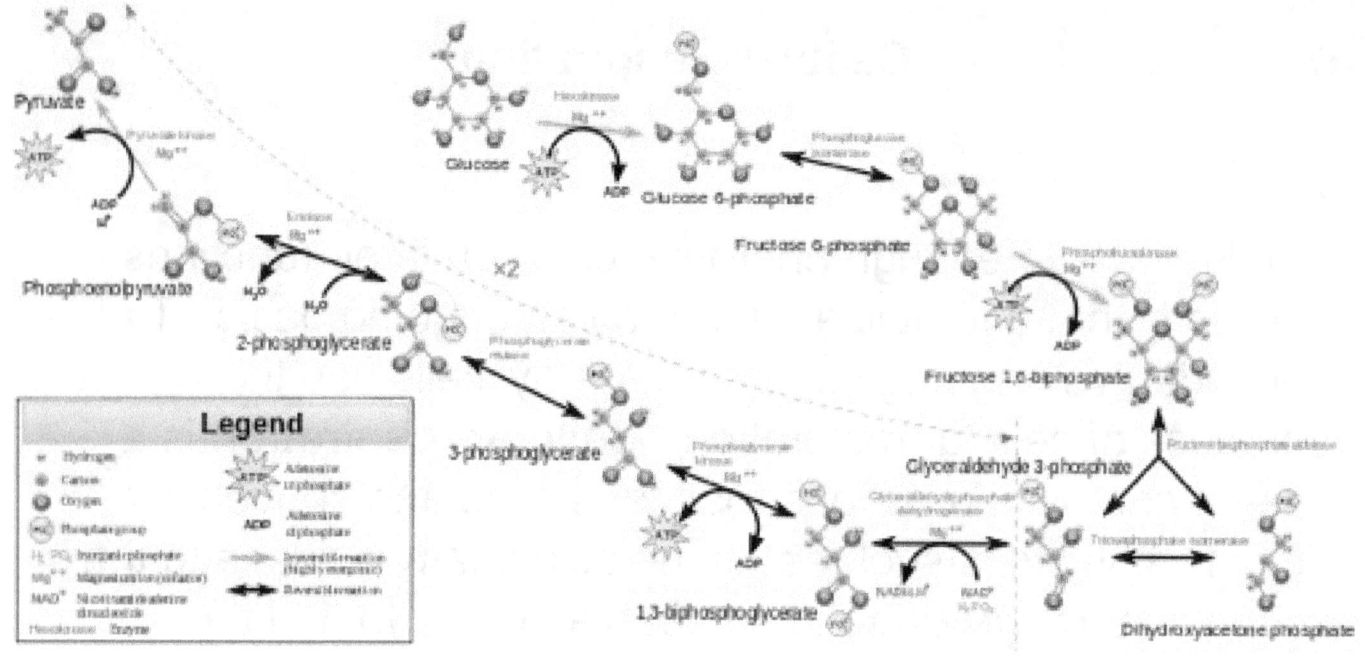

The energy that is released in this reaction is used to form **adenosine triphosphate (ATP)** and NDH. Glycolysis is a process that involves ten different reactions that are all driven by **enzymes** or special proteins that act as catalysts and lower the activation energy of chemical reactions. Glycolysis occurs within the **cytosol** or liquid portion of the cell.

Glycolysis occurs in nearly all organisms on earth. It is referred to by scientists as an ancient metabolic pathway that was used by organisms that existed within the ocean during the Archean eon 2.5 billion years ago. The environment at this time was hot, volcanic, and lacked oxygen. The process of glycolysis during this time was catalyzed by metals and in the presence of enzymes within the aqueous environment of the ocean.

Source:
http://en.wikipedia.org/wiki/Cellular_respiration

Knowledge and Comprehension

Cellular Respiration:

Glucose:

Catabolize:

Glycolysis :

Oxidizing Agent:

Adenosine Triphosphate (ATP):

Pyruvate:

Enzyme:

Cytosol:

1. What is cellular respiration?

2. What is glycolysis?

3. What element must be present in order for glucose to be broken down into pyruvate?

Application, Analysis, Evaluation and Synthesis

4. What is the role of ATP in glycolysis? Explain why it is important to the cell.

5. Explain what an enzyme is and why they are important.

6. Explain how the process of glycolysis first developed on Earth.

Common Core Biology:

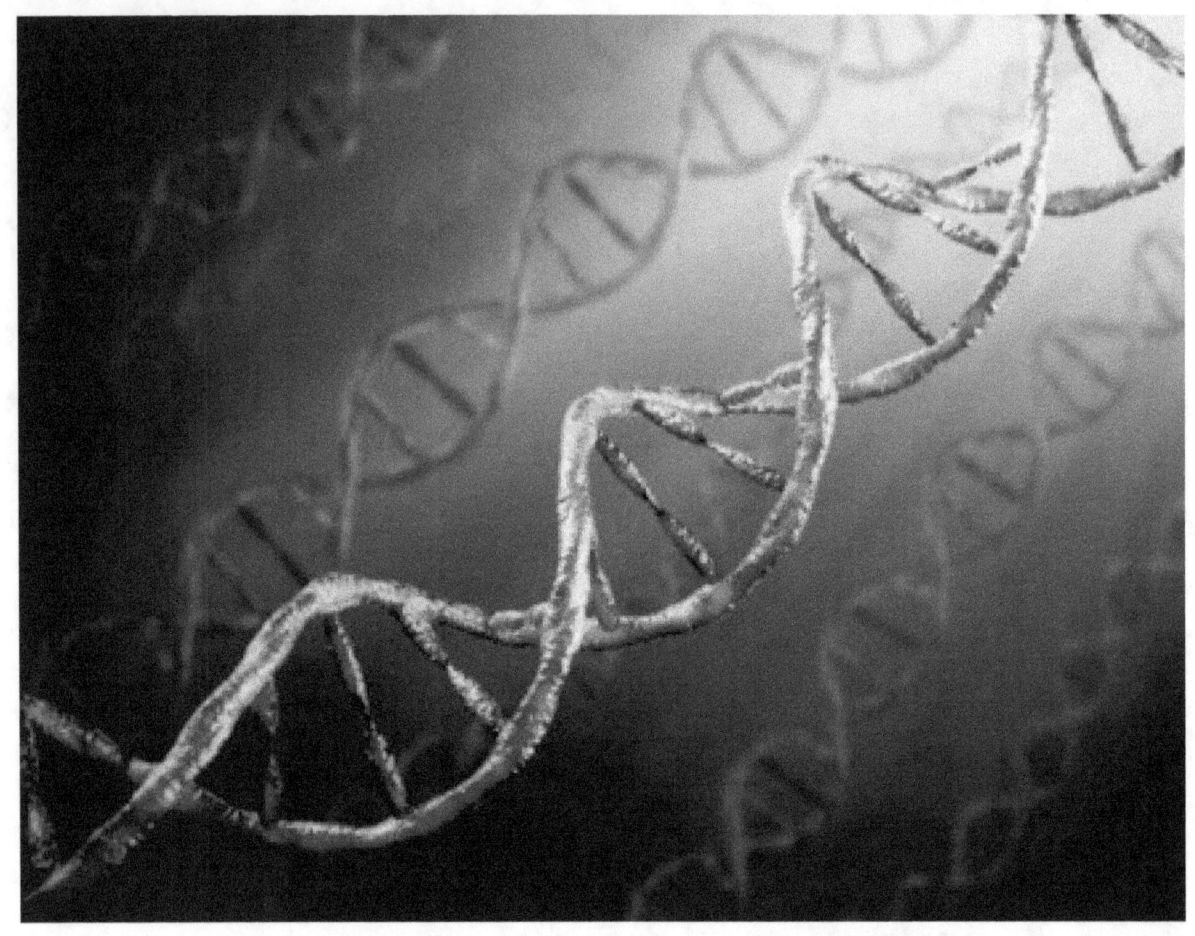

Genetic Information

Monica Sevilla

Contents

The Basic Molecules of Life
What is DNA?
What is RNA?
What are proteins?
What is an amino Acid?
What is a gene?
Gene Regulation
What is a Genetic Mutation?
Transcription
Translation

The Basic Molecules for Life

The cells within organisms use atoms as building blocks for making molecules. **Atoms** are the basic units of elements that can not be divided or broken down. **Molecules** are atoms that are bonded together. Most molecules within living things are made up of the atoms of six different types of elements: sulfur, nitrogen, potassium, hydrogen, oxygen, and carbon. These atoms combine together in different arrangements to forms macromolecules.

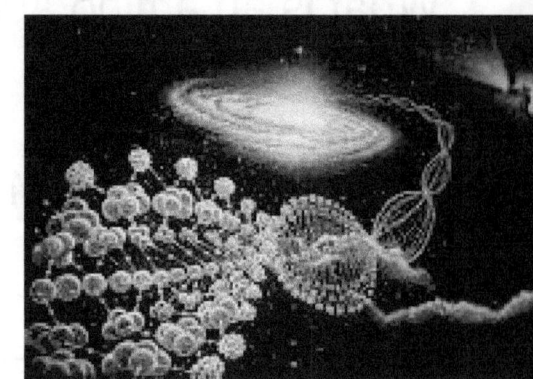

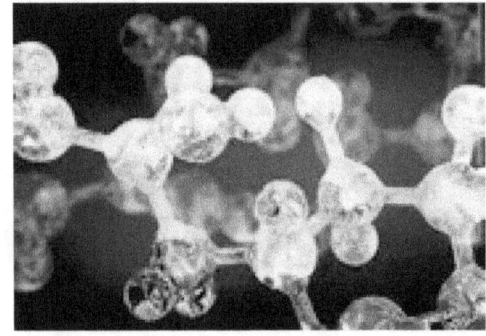

Macromolecules are larger molecules, usually made up of chains of smaller molecules, that the cells either synthesize or make themselves or breakdown and use.

Macromolecules allow the cells to function properly, carry out important chemical reactions, and reproduce. Macromolecules are also classified into four different types: carbohydrates, lipids, nucleic acids, and proteins.

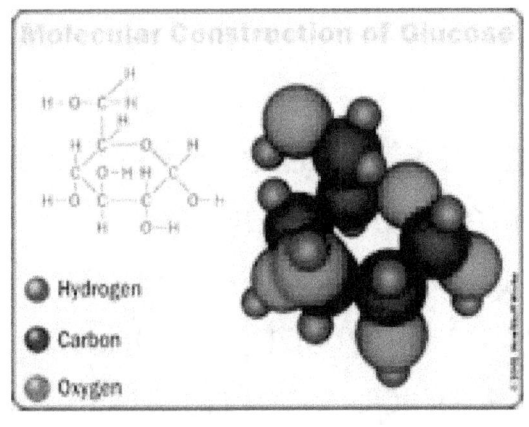

Carbohydrates: are sugar molecules that store energy. When carbohydrates are broken down by the cell, the bonds between the atoms break, and release energy that is used by the cell to fuel different functions.

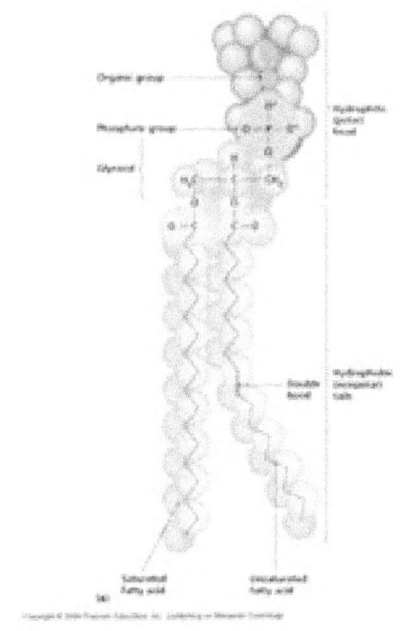

Lipids: are fats that do not dissolve in water. Lipids are classified into different types that include oils, phospholipids, steroids and waxes. Some lipids form barriers against microbes and viruses, prevent loss of water, are used to make hormones, and store large amounts of chemical energy that can be used in the future.

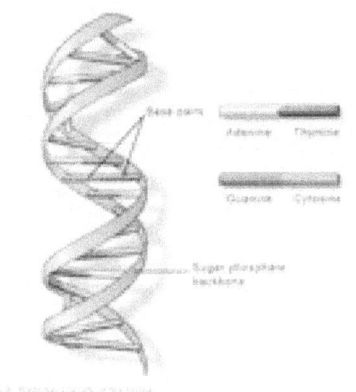

Nucleic Acids: are long chains of molecules called nucleotides. Deoxyribonucleic acid made up of 4 nitrogenous bases: adenine, thymine, guanine, and cytosine. DNA is the genetic material contained within the nucleus of the cell.

**Proteins**: Proteins are long chains of amino acids that have been folded to create a compact structure. They are made from instructions that are coded on the DNA molecule. There are thousands of different proteins, and each has a unique amino acid sequence. They have many different functions and act as catalysts for chemical reactions that occur within organisms.

Knowledge and Comprehension
Words to Know:

Atoms:

Molecules:

Macromolecules:

Carbohydrates:

Lipids:

Nucleic Acids:

Proteins:

1. What is a macromolecule?

2. What four important macromolecules found in organisms.

Application, Analysis, Evaluation and Synthesis

3. What is the difference between a molecule and a macromolecule? What is the relationship between the two?

4. Describe what a nucleic acid is and what a protein is. What is the relationship between nucleic acids and proteins?

5. What do you think would happen to a protein if there was a mistake in the nucleic acid that serves as the genetic instructions for its construction?

6. Identify what the functions lipids are. Why are these functions important?

What is DNA?

DNA (Deoxyribonucleic acid) is a chemical molecule which contains the instructions for the development and functioning of living organisms and viruses on Earth. A **molecule** is the smallest unit of a substance molecule that can exist on its own. Molecules are made up of two or more atoms bonded to each other. The DNA molecule takes the form of a twisted double helix that looks much like a spiral ladder.

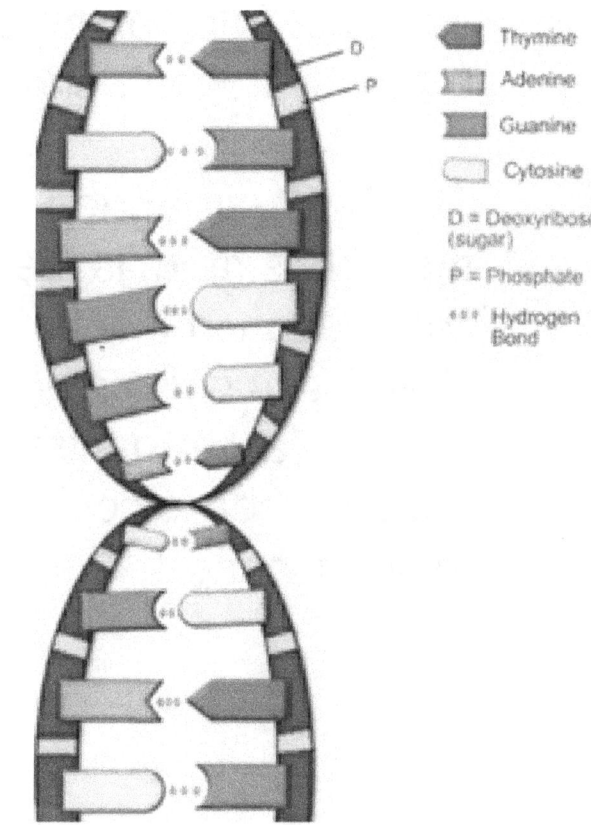

DNA is composed of, sugars called **deoxyribose**, the element phosphorus, and nitrogenous bases. The sugars and phosphorus units are joined together to make up 2 strands of DNA. The **strands** serve as the framework for the molecule. Each strand has different bases that are bonded to each of the deoxyribose sugars on the strands.
There are 4 specific bases.

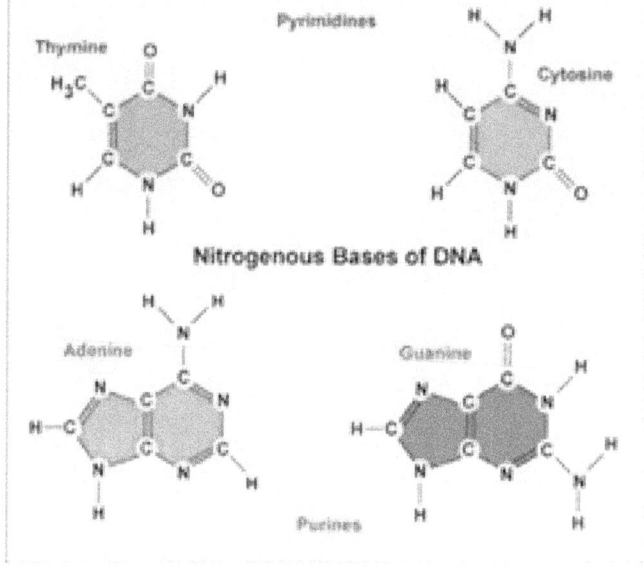

Each of the bases include the element nitrogen. These bases include: adenine, guanine, thymine, and cytosine. Adenine pairs with thymine, and guanine only pairs with cytosine. Each of the pairs of bases are bonded together with a hydrogen bond.

The **DNA sequence**, or arrangement of the bases along the DNA strand, is the genetic code or sequence of information that determines the characteristics, traits, chemical functions and proteins of each organism on Earth. Each organism has a unique DNA sequence. This genetic code is copied and passed down from parents to their offspring.

Knowledge and Comprehension
Words to Know:

DNA:

Molecule:

Deoxyribose:

Strands:

DNA Sequence:

1. What is DNA?

2. What are the 4 bases present in DNA?

Application, Analysis, Evaluation and Synthesis

3. Describe the structure of the DNA molecule.

4. Draw a DNA molecule with 2 nucleotides (4 bases).
   How many different combinations can you make?

Draw a DNA molecule with 3 nucleotides (6 bases)
How many different combinations can you make?

5. If you take into account the 46 chromosomes that humans have, and that each chromosome has a DNA sequence thousands of base pairs long, what can you say about the number of combinations that are possible?

What is RNA?

RNA, or ribonucleic acid, is a **nucleic acid** is a versatile macromolecule that has a variety of roles. Some of these roles include coding for proteins, decoding, regulating the expression of genes, and regulating metabolic reactions. **Messenger RNA** or m-RNA is assembled into a single-stranded chain of nucleotides that is coded for by DNA or deoxyribonucleic acid. A nucleotide is made up of ribose (a sugar), a phosphate group, and a nitrogenous base. The **nitrogenous bases**, bases containing nitrogen, that make up RNA are adenine, cytosine, guanine, and uracil.

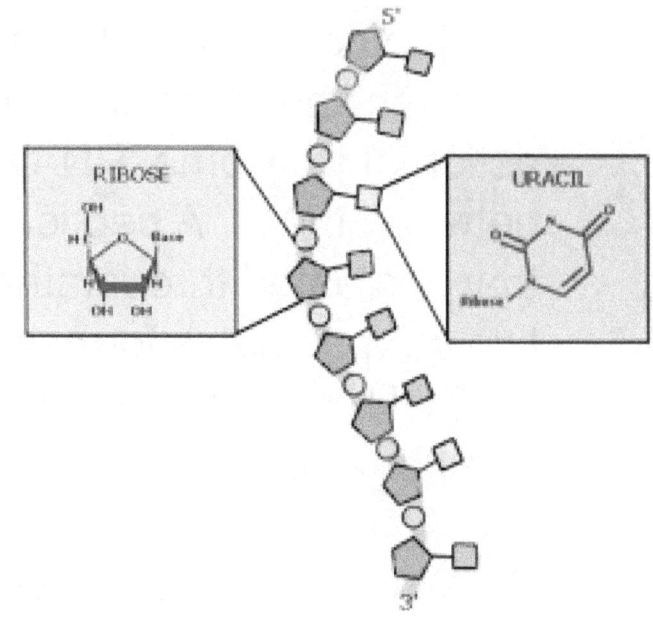

The base **Uracil** is substituted for thymine, which appears in DNA. Uracil is the un-methylated form of thymine. This means that the molecule does not have a methyl group (-CH3), which is one carbon atom bonded to 3 hydrogen atoms.

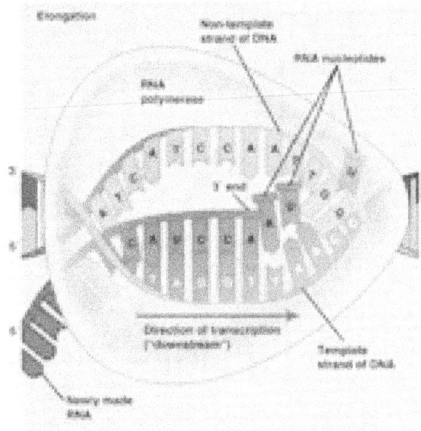

Messenger RNA is used as a template for making proteins within the ribosome of the ribosome of the cell. The ribosome is a special protein complex that directs the assembly of amino acid chains by placing them upon the mRNA itself. Amino acids are selected for and transported by **transfer RNA** or t-RNA. Transfer RNA uses mRNA as a template for placing the amino acids into the right order. The amino acids are then joined together through the action of ribosomal RNA helps to produce peptide bonds between them. A peptide bond is a covalent bond, a bond that is created through the sharing of electrons between two atoms, that is formed between the carboxyl group (-COOH) of one amino acid and the amine group (NH_2) of another amino acid.

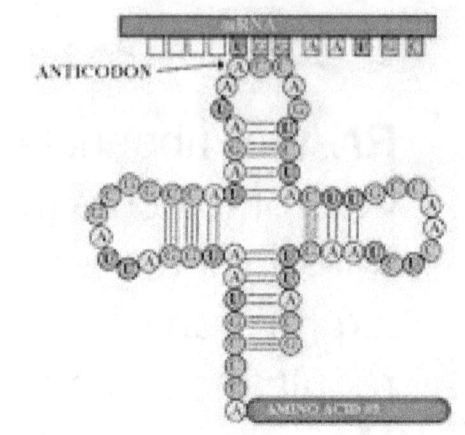

RNA molecules also have a very important role in regulating metabolic reactions within the cell. In its double-stranded form, it folds back on itself, and can act as a catalyst. A **catalyst** speeds up a reaction by lowering its **activation energy** or the energy that is needed to start a reaction. As a result, the rate or the speed of the reaction itself increases at the same concentrations of reactants and at the same temperature. 60% of the ribosome is ribosomal RNA.

Sources:

http://en.wikipedia.org/wiki/RNA

Knowledge and Comprehension:

RNA:

Nucleic Acid:

Messenger RNA:

Nitrogenous Bases:

Uracil:

Transfer RNA:

Ribosomal RNA:

Peptide Bond:

Catalyst:

Activation Energy:

1. What is RNA?

2. What are the functions of RNA?

3. What is a catalyst? How does it help to speed up a reaction?

Application, Analysis, Evaluation, and Synthesis

4. Determine the difference between m-RNA, t-RNA and Ribosomal RNA?

5. Which RNA base is different than what exists in DNA? Explain how this base is different.

What are Proteins?

A **protein** is a complex **polypeptide molecule**, or type of macromolecule made from amino acids that have been linked together in a chain by peptide bonds. A protein is also a product of a **gene** that is encoded by a specific segment of the DNA and have a special job or function within the cell. Proteins can have many different functions or jobs in the cell. Some of these functions include: copying DNA, synthesizing macromolecules such as amino acid chains, folding amino acid chains, breaking down macromolecules, transporting molecules, regulating transcription, controlling gene expression, catalyzing metabolic reactions, and responding to stimuli in the environment.

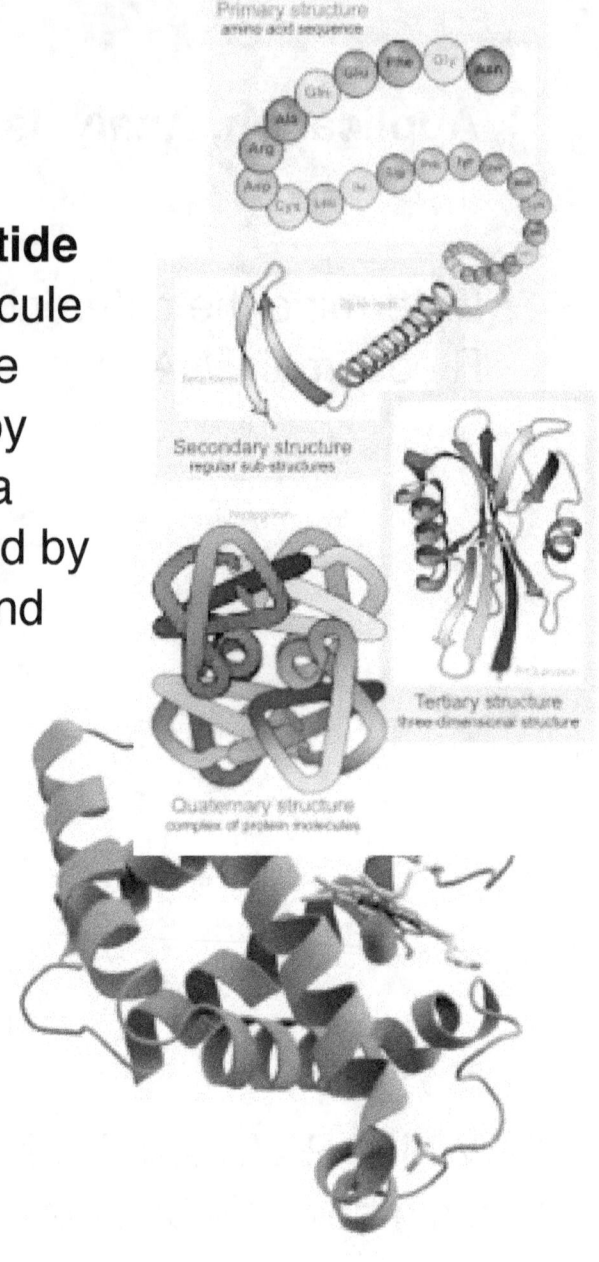

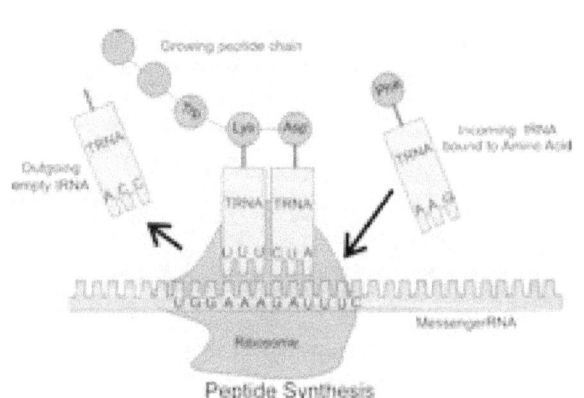

The polypeptide chain is synthesized or made through the **transcription** of DNA or the copying of DNA into m-RNA. **Translation** of the polypeptide chain occurs as transfer RNA (t-RNA) selects and carries amino acids to the ribosome and are placed onto the m-RNA strand for assembly. Peptide bonds are formed and link each of the amino acids into a chain. The chain is released when the ribosome encounters a stop codon on the m-RNA strand.

Proteins gain their 3-dimensional structure through the natural folding of the amino acid chain or with the assistance of chaperone proteins or enzymes which help the chain to fold at the correct locations. Hydrogen bonds are then created between specific molecules to stablize and maintain the shape of the protein. The final structure and shape of a protein determines its function and its activity. Proteins that have been synthesized from DNA with mutations or changes will sometimes code for damaged proteins that do not have the correct shape and cannot function properly.

Protein complexes can be created by the bonding of more than one amino acid chain together. These amino acids are coded for by different genes and work together as one unit to perform a job or function. A classic example of a well-studied protein complex is the **ribosome**, the protein complex where m-RNA is translated into an amino acid chain. The ribosome is a large protein complex known as a ribonucleoprotein which contains two functional protein subunits: a ribosomal subunit that reads the mRNA sequence and a large subunit that joins amino acids together into a polypeptide chain.

Knowledge and Comprehension:

Protein:

Polypeptide:

Gene:

Transcription:

Translation:

Protein Complex:

Ribosome:

1. What is a protein?

2. Describe the important functions that proteins have.

3. What can happen to the function of a protein its gene has mutations in it?

Application, Analysis, Evaluation, and Synthesis

4. Explain how a protein is synthesized or made.

5. Explain why the ribosome is an example of a protein complex.

6. Explain why the function of a protein depends on its structure and shape. What can affect its shape?

What is an Amino Acid?

Amino acids are important **macromolecules**, or complex molecules, that serve as the building blocks for proteins. Amino acids have a basic chemical structure that is linear and that contains a carboxyl group (-COOH), an amine group ($-NH_2$) and an R-group which is a side chain that is specific to the amino acid. The common elements for life, carbon, oxygen, nitrogen and hydrogen, exist within amino acids. 500 amino acids are known to exist.

20 **essential amino acids** exist on Earth. They are called "essential" because these amino acids cannot be made by humans and organisms and naturally exist. Humans and organisms must ingest or take in these macromolecules into the body in the form of nutrients. These nutrients can be found in foods, drinks, or taken as a supplement such as a

Essential amino acids	Non-essential amino acids
Histidine	Alanine
Isoleucine	Arginine
Leucine	Asparagine
Lysine	Aspartic acid
Methionine	Cysteine
Phenylalanine	Glutamic acid
Threonine	Glutamine
Tryptophan	Glycine
Valine	Proline
	Serine
	Tyrosine

vitamin. These 20 essential amino acids are used to make proteins and involved in different metabolic functions.

A **protein** is a complex macromolecule made from amino acids that have been linked together in a chain by peptide bonds. The polypeptide chain is synthesized or made through the **transcription** of DNA or the copying of DNA into m-RNA. **Translation** of the polypeptide chain occurs as transfer RNA (t-RNA) selects and carries amino acids to the ribosome and are placed onto the m-RNA strand for assembly. Peptide bonds are formed and link each of the amino acids into a chain.

Non-essential amino acids are amino acids that are not essential or required by the body to synthesize or make proteins and other macromolecules. Essential amino acids can be found in enzymes, hormones, transport molecules, immune system molecules, neurotransmitters, and structural proteins. Non-essential amino acids can be made by the body from other molecules that already exist.

Biosynthesis of Non-essential Amino Acids

Amino Acid	Source
Alanine	pyruvate
Aspartic acid	oxaloacetate
Asparagine	oxaloacetate via aspartic acid
Glutamic acid	alpha ketoglutarate
Glutamine	alpha ketoglutarate via glutamic acid
Histidine**	5-aminoimidazole-4-carboxamide ribotide
Proline	alpha ketoglutarate via glutamic acid
Arginine**	from glutamate via ornithine
Serine	3-phosphoglyceric acid
Glycine	3-phosphoglyceric acid via serine
Cysteine*	3-phosphoglycerate via serine and methionine (cystathionine)
Tyrosine*	phenylalanine

The molecules undergo a chemical reaction that changes them into a non-essential amino acid. These 11 amino acids include: as alanine, glutamine, glycine, asparagine, aspartic acid, glutamic acid, cysteine, proline, serine, tyrosine, and arginine.

Knowledge and Comprehension:

Amino Acids:

Macromolecules:

Essential Amino Acids:

Protein:

Transcription:

Translation:

Non-Essential Amino Acids:

1. What is an amino acid?

2. Describe the basic structure of an amino acid.

3. What are essential amino acids?

Application, Analysis, Evaluation, and Synthesis

4. Explain the relationship between amino acids and proteins.

5. Explain how proteins are made with amino acids.

6. Explain how essential amino acids and non-essential amino acids are different.

What is a Gene?

A **gene** is a segment of DNA that contains the genetic information or instructions for a trait, a polypeptide, or for making specific proteins within the cell. The word gene comes from the Greek word "genos" which means birth or origin. It is the basic molecular unit of heredity that is located on the chromosomes of living organisms. **Chromosomes** contain many genes, other sequences of nucleotides, and regulators, segments of DNA that slow down or speed up the expression of RNA and proteins of specific genes. Chromosomes also contain proteins that are bound to the DNA molecule. These proteins package DNA and and control the function of the DNA itself. Most of the DNA on the chromosome codes for an organism's genetic information. Some of the DNA on the chromosome is **non-coding**, do not code for genetic information, and direct DNA replication proteins to start and stop activity.

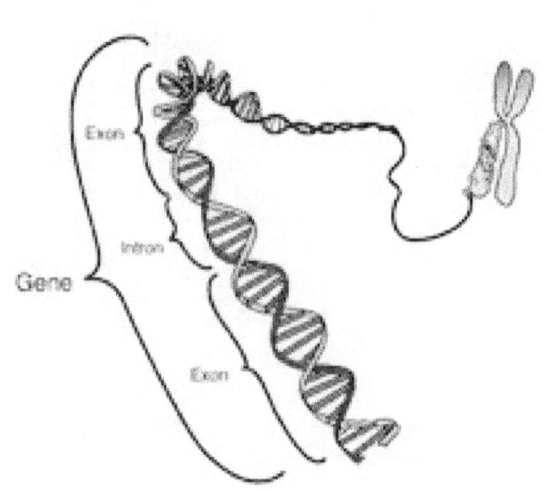

In **eukaryotes**, DNA is coiled around special proteins called **histones** that compact and condense the DNA into a smaller amount of space. These chromosomes reside

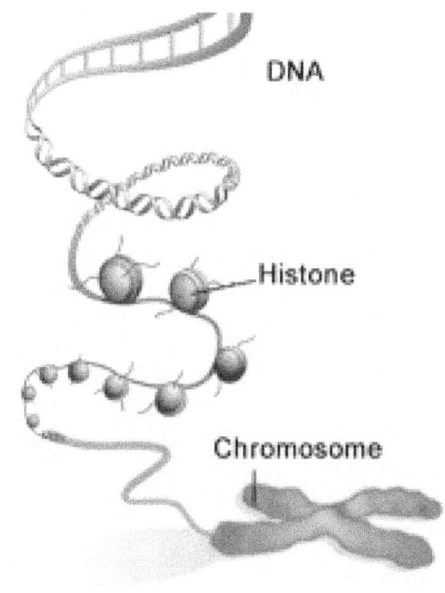

within the nucleus of cells. In **prokaryotes** that do not have a nucleus, chromosomes exist as circular units within the cells.

Some genes have **alleles** which are variants or different expressions of a gene. The expressions of these genes correspond to the phenotypes, observable characteristics or physical traits of an organism. Alleles can be dominant or recessive. Dominant alleles are expressed more often than recessive alleles. An example of this is eye color in humans. The different colors are expressions of different alleles with the genes. The brown eye color is dominant. The blue eye color is recessive. The brown eye color is expressed more often than the blue eye color.

Knowledge and Comprehension
Words to Know:

Genes:

Chromosomes:

Non-coding:

Eukaryotes:

Prokaryotes:

histones:

1. What is a gene?

2. How is the gene related to the chromosome?

3. What do chromosomes contain besides genes?

Application, Analysis, Evaluation and Synthesis

4. What is the difference between the DNA in a eukaryote and DNA in prokaryote.

5. What is an allele and how do different expressions of the same trait occur? What is an example of this?

Gene Regulation

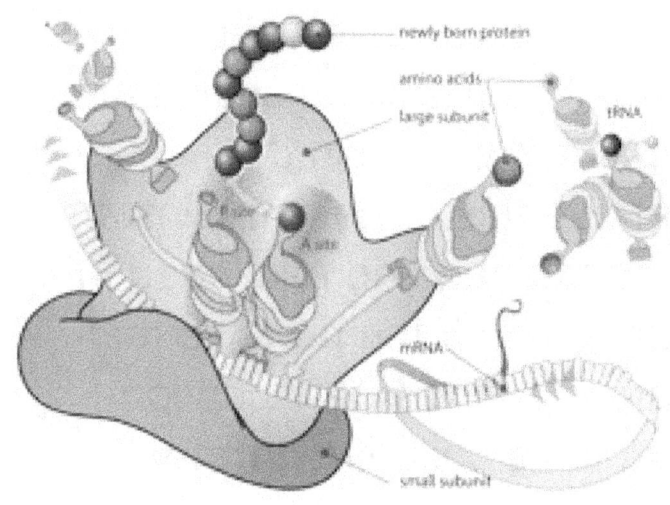

Changes in cells are the result of the expression of certain genes within the DNA of each cell. **Gene expression** is the process by which information is used to make a product that has a defined **function** or job such as a protein or an enzyme. The different types of cells will "turn on" some of its genes. The other genes are repressed or "turned off." During the development of an organism from conception to adulthood, genes are turned on and off in patterns. This is what is known as **gene regulation**.

One of the most important functions of gene regulation is the ability of the genes to respond to changes within the environment. Changes in the chemical reactions within the cell can affect genes by turning them on or off. Gene regulation mostly occurs during the process of **transcription** when RNA is synthesized or made from DNA. This process uses the information that is present within the gene which is a section of DNA that codes for a product

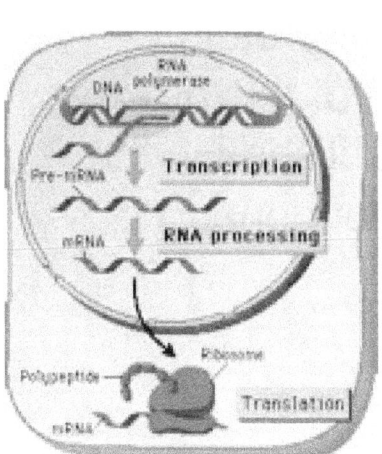

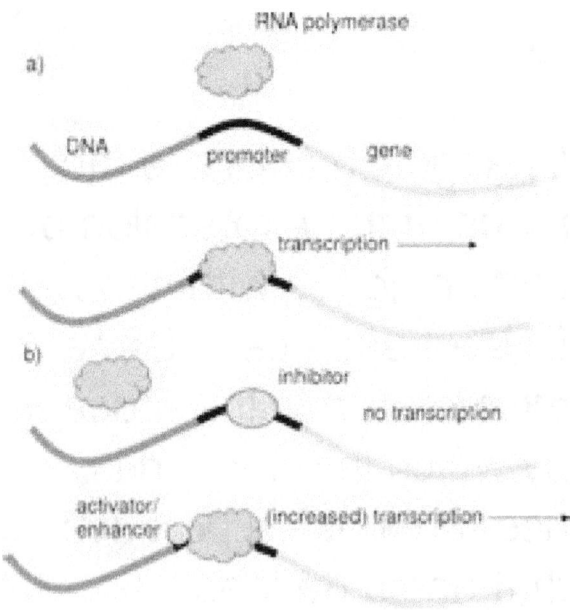

such as a protein, to make messenger RNA. The messenger RNA later enters the process of translation where amino acid chains are assembled within the ribosomes. These chains are then folded and packaged into proteins. If the proteins that are made are transcription proteins called **promoters** and will initiate or start the process of transcription by binding to the DNA within the cell. **Inhibitors** stop the process of transcription by binding to the promoters and changing their shape so that they essentially fall off the DNA they were attached to.

Promoters and inhibitors control and balance the level of proteins that are made by a gene by controlling the process of transcription. When the concentration of a specific protein is low, more promoters are made to turn on transcription. When the concentration of protein is high, more inhibitors are made. The result of this process causes equilibrium or a balance to be established in the production of proteins and the expression of genes.

Knowledge and Comprehension:

Gene Expression:

Function:

Gene Regulation:

Transcription:

Promoters:

Inhibitors:

1. What is gene expression?

2. What is gene regulation?

Application, Analysis, Evaluation, and Synthesis

3. Explain how gene regulation works.

4. How are promoters and inhibitors different?

5. Why are promoters and inhibitors important for regulating the expression of genes?

What is a Genetic Mutation?

A genetic **mutation** is a change in the DNA nucleotide sequence of a gene. A **nucleotide** is made up of a nitrogenous base, the sugar deoxyribose, and a phosphorus atom. In living organisms, many of the genes code for proteins. They are the genetic instructions for the structure and the function of these proteins.

When a genetic mutation occurs, the nucleotide sequence can be changed by the **insertion** (addition of bases) or **deletion** (eliminating bases) of bases. These "mistakes" in the DNA sequence can cause changes to the structure of the protein which can effect the function or the activity of the protein the gene codes for. Mutations may be caused by mistakes during the replication of DNA or damaged that has occurred by mutagens in the environment such as ultraviolet radiation or toxic chemicals.

Many mutations are not harmful. These mutations sometimes code for a trait or characteristic that is improved. Sometimes no change occurs. Some mutations, however may affect the survival of living things. One important example of this are mutations that occur to the DNA of some genes that cause cancer. Cancer is the uncontrolled growth of cells into tumors. The genes that code for the **BRCA proteins,** DNA repair proteins, will sometimes carry a mutation that impacts its ability to bind to damaged DNA within the nucleus of the cell.

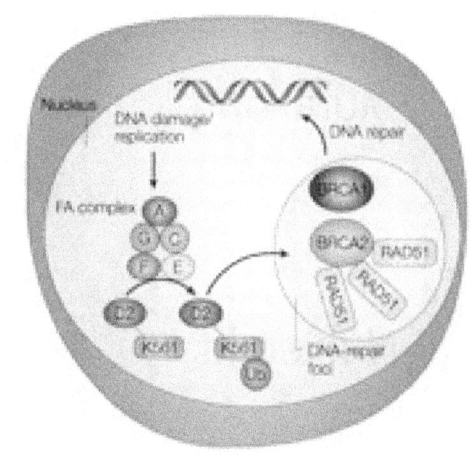

Nature Reviews | Cancer

The change that occurs in the protein's structure causes the protein to lose its function or ability to repair DNA. This protein is very important. Without its proper function, damaged DNA accumulates and may cause the cell to reproduce in an uncontrolled way, leading to breast cancer and cancer of the ovaries.

Knowledge and Comprehension
Words to Know:

Mutation:

Nucleotide:

Insertion:

Deletion:

BRCA proteins:

1. What is a mutation?

2. What is a nucleotide made up of?

Application, Analysis, Evaluation and Synthesis

3. Explain how a mutation in the DNA can occur.

4. Explain what the difference is between an insertion and a deletion?

5. Are all mutations in the DNA harmful? Why or why not?

6. Explain how breast cancer and cancer of the ovaries occurs. Use evidence from the text to support your answer.

Transcription

Transcription is the process in which DNA within the nucleus of the cell is copied into messenger RNA (m-RNA) used by the ribosomes to produce proteins. DNA is used as a template to produce a different nucleic acid that is complementary to DNA. This occurs through the action of the enzyme RNA polymerase. An **enzyme** is a special protein that carries out a specific function or job. In the case of RNA polymerase, **messenger RNA (m-RNA)** is produced by **polymerizing** or attaching **ribonucleotides** or RNA bases, together into a chain. The ribonucleotides include adenine, guanine, cytosine, and uracil. Uracil is used instead of thymine. Polymerization of the RNA begins at the 3' end of the RNA chain and proceeds toward the 5' end.

Control of transcription affects the expression of genes. Genes can be turned "on" or "off" by controlling the activity of DNA polymerase. Transcription is initiated at locations on the DNA called **promoters**. If an **inhibitor** is preset and binds to DNA polymerase, the shape of the enzyme changes and it falls off the DNA strand. Replication is terminated at locations on the DNA called **terminators**. The expression of the gene is turned "off" or inhibited and its gene product, a protein, is not produced. This mechanism is used as part of a communication and feedback system where the cell can respond to different conditions within the environment, and be able to adapt to the environment by producing more or producing less proteins.

Knowledge and Comprehension:

Transcription:

Enzyme:

Messenger RNA (m-RNA):

Polymerize:

Ribonucleotides:

Promoter:

Inhibitor:

Terminator:

1. Describe what transcription is.

2. Indicate what the function of an enzyme is.

3. Describe what occurs when a ribonucleotide or RNA base is polymerized.

Application, Analysis, Evaluation, and Synthesis

4. Explain why RNA polymerize is important.

5. Explain how gene expression is controlled through the process of transcription.

Translation

Translation is the process in which RNA is transformed into an amino acid chain or a **polypeptide chain**. Translation occurs through the action of the ribosome.

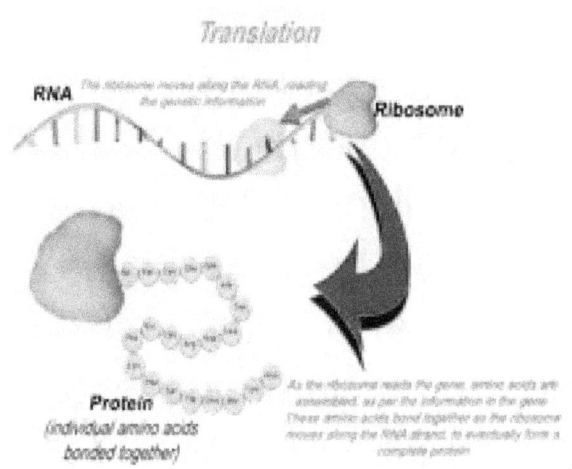

The **ribosome** is a large complex known as a ribonucleoprotein which contains two functional protein subunits: a ribosomal subunit that reads the mRNA sequence and a large subunit that links **amino acids**, the basic building blocks of proteins, together into a polypeptide chain. These two subunits work together as one unit, a complex protein complex. The ribosome is located within the cytosol or fluid part of the cell.

Ribosomes start the translation process by binding to **messenger RNA (m-RNA)**. m-RNA serves as a template in which amino acids are placed. The amino acids needed to build the polypeptide chain are selected and carried to the

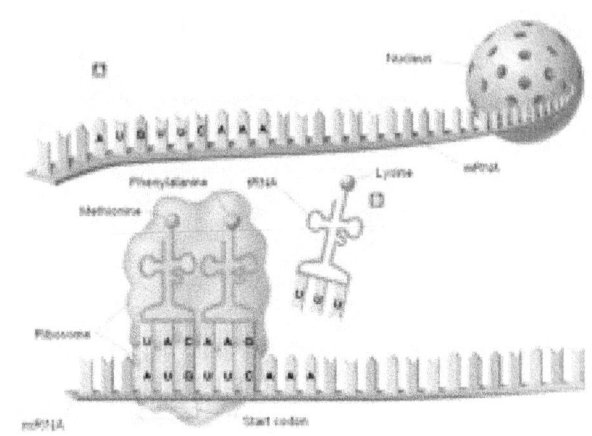

ribosome by **transfer RNA (t-RNA).** The t-RNA binds directly to the m-RNA chain. Each amino acid is represented by a specific sequence of 3 ribonucleotides or RNA bases. Peptide bonds are formed between each amino acid within the ribosome creating a polypeptide or many peptide chain. The polypeptide is terminated when the ribosome encounters a stop codon. This codon induces the binding of a protein know as a release factor. The **release factor** promotes the destruction and disassembly of the ribosome. The resulting polypeptide chain is released and ready for transport by the vesicles. folded into a functional protein by special proteins. mRNA is used as a template for the assembly of an amino acid or polypeptide chain. RNA is complementary to DNA and is encoded by the DNA within the genes.

Knowledge and Comprehension:

Translation:

Polypeptide Chain:

Ribosome:

Amino Acids:

Messenger RNA (m-RNA):

Transfer RNA (m-RNA):

Release Factor:

1. What is translation?

2. Why is transfer RNA (t-RNA) needed for the assembly of amino acid chains?

Application, Analysis, Evaluation, and Synthesis

3. What is a ribosome? What is its function? Why is it referred to as a protein complex?

4. Describe the process of translation. How is an amino acid chain assembled in the ribosome?

5. What is the function of m-RNA? What is the function of t-RNA? How do they work together to make an amino acid chain?

The Structure of the Cell

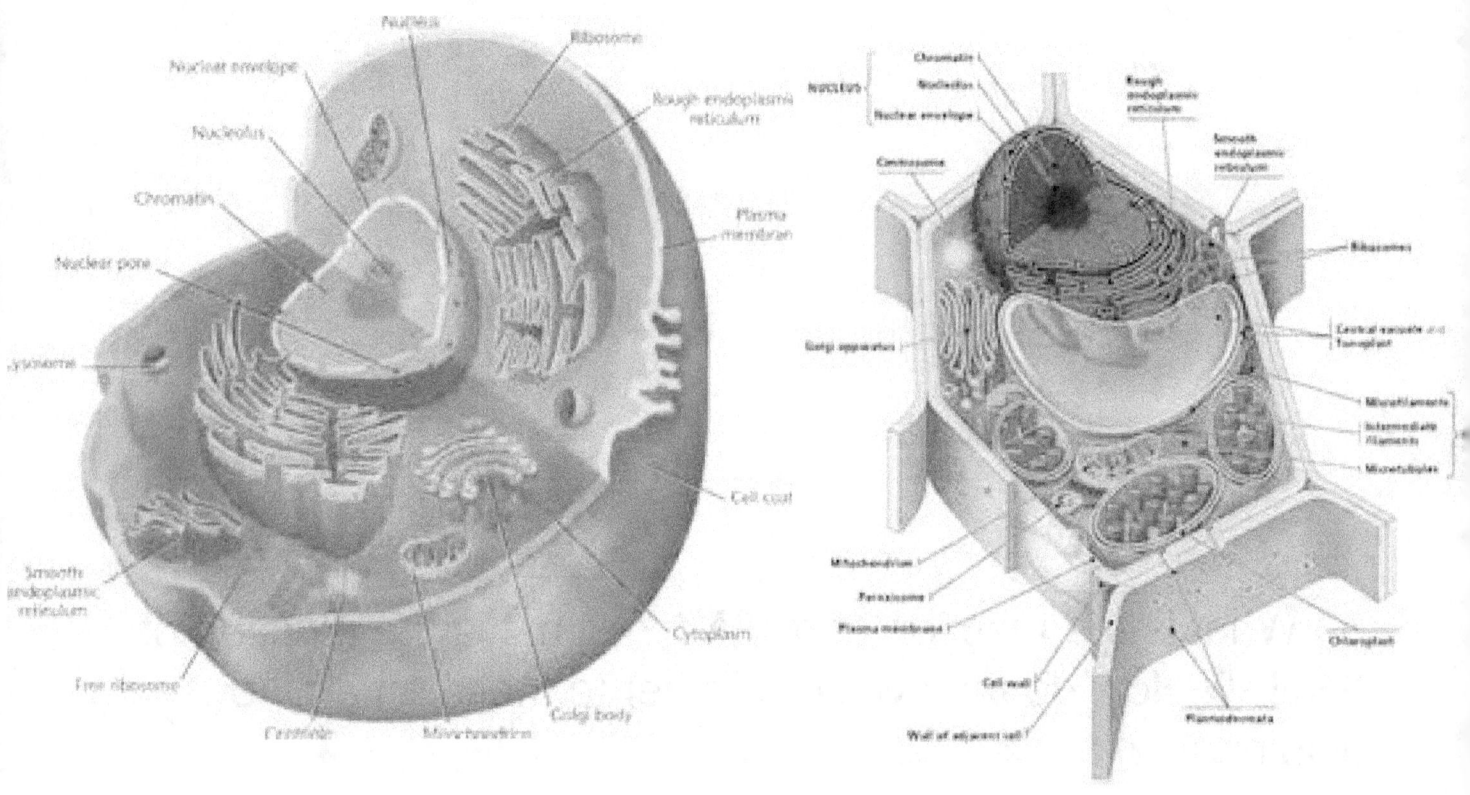

Common Core Workbook

Monica Sevilla

Contents

What is a Cell?
Types of Cells
The Plant Cell
Parts of the Cell
The Cell Membrane
The Nucleus of the Cell
The Mitochondrion
The Endoplasmic Reticulum
Golgi Apparatus
The Cell Organelles and their Functions
Study Cards
The Structure of Animal and Plant Cells
Activity
The Cell Unit Exam

What is a Cell?

A **cell** is the smallest functional and structural building block of life. Cells make up all living things. They communicate with each other and work together to make up the structure of living things and to carry out the many functions of organism.

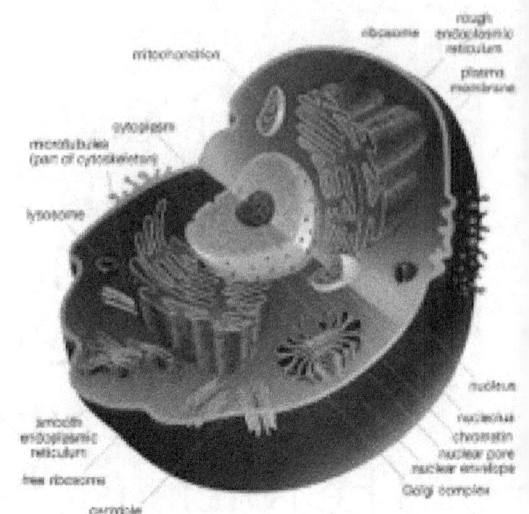

The first person to describe the cell from his observations was **Robert Hooke** in the 1660's. In 1665, he built a microscope to look at objects that could not be seen with the naked eye. A **microscope** is a scientific tool that use glass lenses to magnify tiny objects. The first observation he made was looking at a sample of dead and dry bark from a cork tree. What he saw were many little rooms or "cells" that were held closely together.

The next observation Hooke made was looking at a slice of a living plant. He also saw cells in this sample as well. He noticed that these cells had fluid inside them and called them "juicy." He also notice that plants cells have a **cell wall** or the outermost structure of

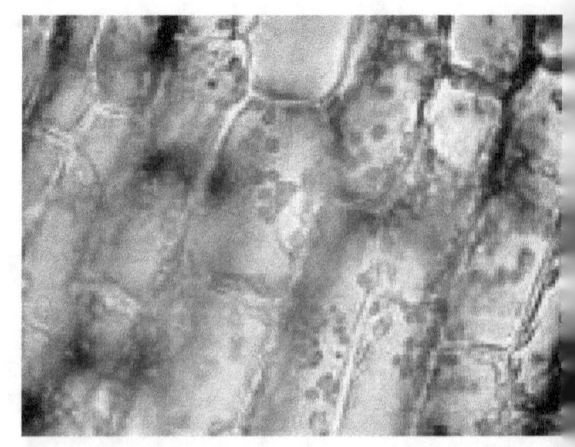

the cell.

Anton van Leeuwenhook, in 1673, built and used his own microscope to look at a sample of scum that he found in a pond. He noticed that there were small organisms swimming in the water. He called these organisms 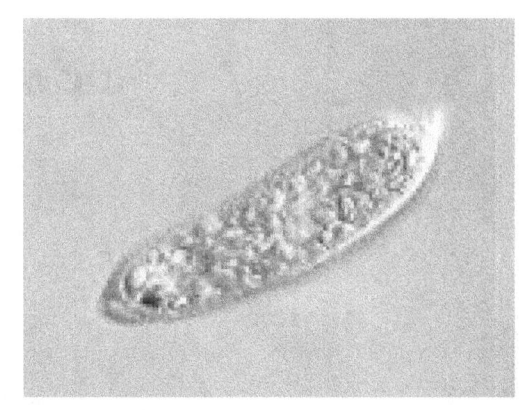 "little animals" or animalcules. We now know these organisms as **protists** or single celled animals. He was also the first person to use his microscope to see bacteria, a type of single celled animal.

Knowledge and Comprehension
Words to Know:

Cell:

Robert Hooke:

Cell Wall:

Anton van Leeuwenhook:

Protists:

Microscope:

1. What is a microscope?

2. What can you see with a microscope that you can not see with your unaided eye?

Application, Analysis, Evaluation and Synthesis

3. Describe the contribution that Robert Hooke made to science. What did he discover and why was it important.

4. Describe the contribution that Anton van Leeuwenhook made to science. What did he discover and why was it important.

5. Compare and contrast the contributions that were made by Robert Hooke and Anton van Leewenhook. What was similar about their discoveries? What was different?

6. If the discovery of the microscope had not been made, how would this have impacted or affected our world?

What is a Cell?

A **cell** is the smallest functional and structural building block of life. Cells make up all living things. They communicate with each other and work together to make up the structure of living things and to carry out the many functions of organism.

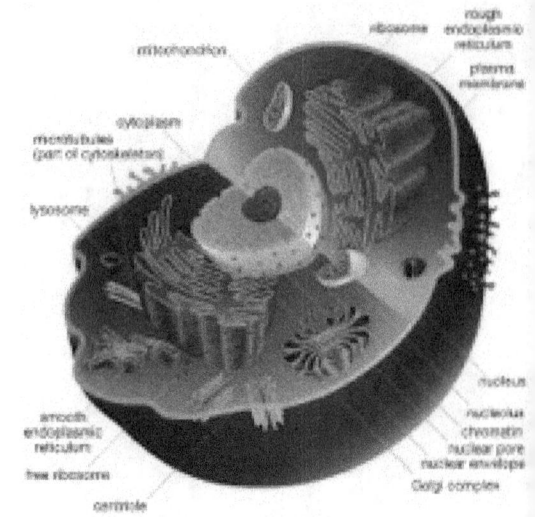

The first person to describe the cell from his observations was **Robert Hooke** in the 1660's. In 1665, he built a microscope to look at objects that could not be seen with the naked eye. A **microscope** is a scientific tool that use glass lenses to magnify tiny objects. The first observation he made was looking at a sample of dead and dry bark from a cork tree. What he saw were many little rooms or "cells" that were held closely together.

The next observation Hooke made was looking at a slice of a living plant. He also saw cells in this sample as well. He noticed that these cells had fluid inside them and called them "juicy." He also notice that plants cells have a **cell wall** or the outermost structure of

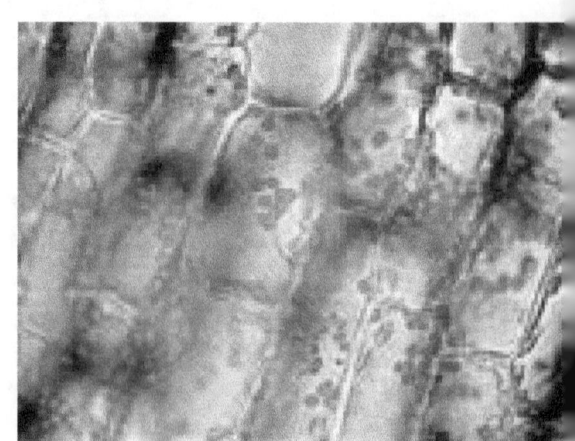

the cell.

Anton van Leeuwenhook, in 1673, built and used his own microscope to look at a sample of scum that he found in a pond. He noticed that there were small organisms swimming in the water. He called these organisms "little animals" or animalcules. We now know these organisms as **protists** or single celled animals. He was also the first person to use his microscope to see bacteria, a type of single celled animal.

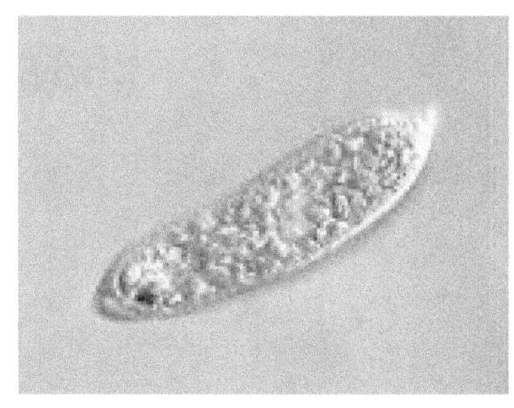

Knowledge and Comprehension
Words to Know:

Cell:

Robert Hooke:

Cell Wall:

Anton van Leeuwenhook:

Protists:

Microscope:

1. What is a microscope?

2. What can you see with a microscope that you can not see with your unaided eye?

Application, Analysis, Evaluation and Synthesis

3. Describe the contribution that Robert Hooke made to science. What did he discover and why was it important.

4. Describe the contribution that Anton van Leeuwenhook made to science. What did he discover and why was it important.

5. Compare and contrast the contributions that were made by Robert Hooke and Anton van Leewenhook. What was similar about their discoveries? What was different?

6. If the discovery of the microscope had not been made, how would this have impacted or affected our world?

Types of Cells

A **cell** is the smallest functional and structural building block of life. Cells make up all living things. They communicate with each other and work together to make up the structure of living things and to carry out the many functions of organism.

There are two different types of cells. Cells that have a nucleus and cells that do not have a nucleus. The **nucleus** is a structure within the cell that stores all the genetic information known as **DNA** for the function and reproduction of the cell.

Eukaryotic cells have a nucleus. They have genetic material DNA. Plant cells and animal cells are examples of eukaryotic cells.

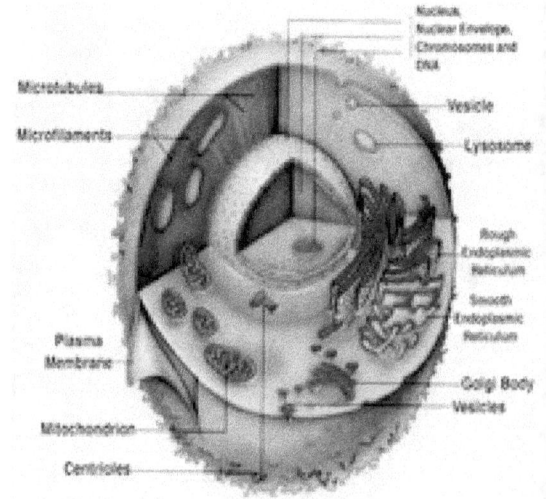

Prokaryotic cells or prokaryote do not have a nucleus. Prokaryotic cells are classified into two groups: bacteria and archaea.

It is believed by scientists that the

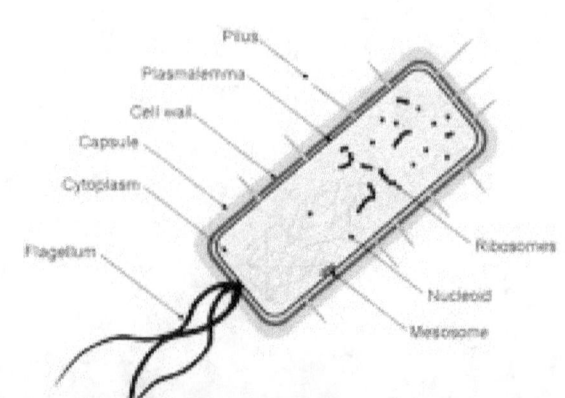

first living things on Earth were probably prokaryotes. More complex life forms are said to have evolved from these single celled organisms. The evidence supports this fact is that microbials, 3.48 billion years old, were discovered in western Australia. Other evidence that supports this is the research that has been done by biochemist Douglas Theobaldis who calculated that the last universal common ancestor of all life forms on Earth is lease 10^{2860} more probable than having multiple ancestors. This evidence suggests that life on Earth followed an evolutionary pattern suggested by Charles Darwin in his book **On the Origin of Species**. In his book, he infers that all living organisms on Earth arose through evolutionary processes from one primordial form.

Sources:

http://news.nationalgeographic.com/news/2010/05/100513-science-evolution-darwin-single-ancestor/

Knowledge and Comprehension
Words to Know:

Cell:

Nucleus:

DNA:

Eukaryotic Cells :

Prokaryotic cells:

1. What is a cell?

2. What are the two major types of cells?

Application, Analysis, Evaluation and Synthesis

3. Explain how eukaryotes and prokaryotes are different from each other. What do these organisms have in common?

4. Could a living organism survive without a nucleus? Find evidence from the text to support your answer.

5. Do you agree with the claim "All complex organisms evolved from a single-celled organism." Find evidence from the text to support your answer.

What is a Plant Cell?

Plants are **eukaryotes** or a multicellular organisms. **Plant cells** have unique features when compared to animal cells. They have most of the organelles and structures that exist within the animal cell, but also have a cell wall, chloroplasts, and a large central vacuole that have specialized functions to carryout the process of photosynthesis. **Photosynthesis** is the process of a plant making the simple sugar glucose with the light energy from the sun.

Plants have organelles called **chloroplasts** which is the location of photosynthesis. The chloroplasts use carbon dioxide and water molecules and convert them glucose through a series of chemical reactions. The chloroplasts contain **thylakoid membranes** which house the green pigment chlorophyll. **Chlorophyll** absorbs or takes in light energy within its molecular bonds. Electrons are then transferred from chlorophyll molecules to reaction centers

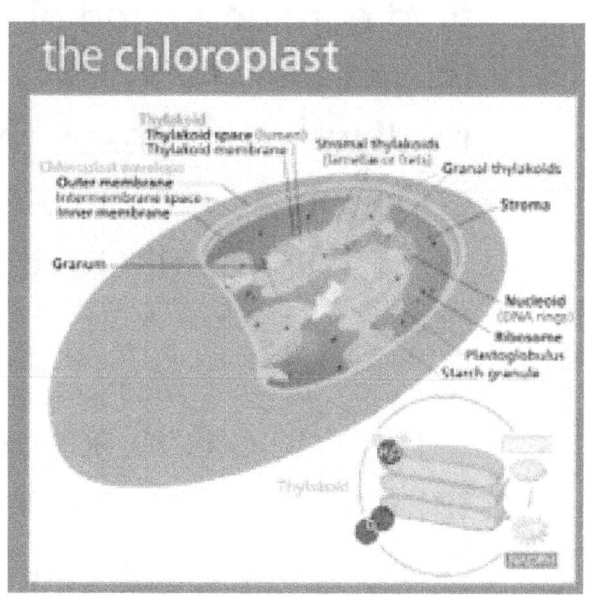

where it is used in photosynthesis.

The **cell wall** is a structure that surrounds the cell membrane of plant cells. The cell wall is rigid. It maintains the shape of the cell and protects the cell fro damage. Cell walls allow fluids and molecules to pass in and out of the cell. Unlike the animal cell, which is made of phospholipids and proteins, the cell wall is made mostly of a carbohydrate called **cellulose** and protein.

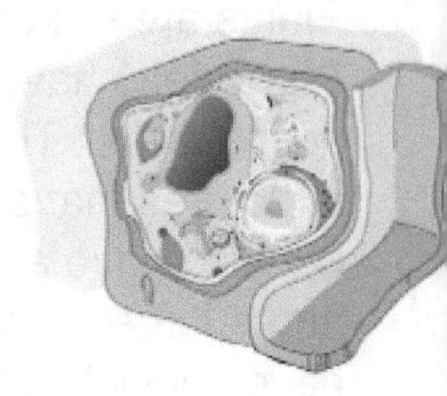

The **large central vacuole** maintains the internal pressure of fluids within the plant cell. This structure is a water-filled membrane that regulates the amount of fluid coming in and out of the cell. The large central vacuole draws water into itself when it senses that the concentration of solutes is low outside cell. This maintains the rigidity of the cell.

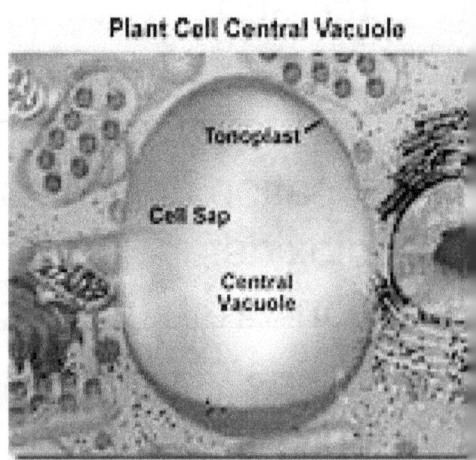

When the solute concentration is higher on the inside of the cell then the outside, the large central vacuole will release water out of the cell in order to decrease the concentration. In the plant cell, a solute is a mixture of

water and other substances such as molecules and minerals. These two actions are how the cells respond to changes in their environment in order to maintain equilibrium. **Equilibrium** is a state where the environments both inside and outside of the cells are in balance. In this case, the cells try to balance the solute concentration, and the pH of the solution through the movement of water in and out of the cells through a semi-permeable membrane. This process is known as **osmosis**.

Knowledge and Comprehension
Words to Know:

Eukaryotes:

Photosynthesis:

Chloroplasts:

Chlorophyll:

Thylakoid membranes:

Cell Wall:

Large Central Vacuole:

Cellulose:

Equilibrium:

Osmosis.

1. What three organelles or structures are unique to the plant cell when compared to an animal cell?

2. What is osmosis?

Application, Analysis, Evaluation and Synthesis

3. What function does the plant cell serve? How are the cell wall in a plant and the outer membrane of the animal cell different?

4. How does the plant cell maintain equilibrium?

5. If the solute concentration is higher in side the cell than it is outside the cell, what must occur for the cell to achieve equilibrium?

6. Predict what would occur if the central vacuole inside the plant cell was damaged? What would happen to the cell?

7. What is a chloroplast and what is its function within the plant cell? Could the cell function without the chloroplast? why or why not?

8. What is chlorophyll and what is its function in the plant cell? What is its role in the process of photosynthesis.

The Cell Membrane

The cell membrane is the protective barrier that surrounds the cell. It is a flexible covering that is made up of one or more layers of phospholipid molecules bonded or linked together. These layers are made of **phosholipids**. Lipids are fat. Phospholipids are made of a phosphate molecule, a glycerol molecule, and 2 fatty acid chains. The animal cell has a cell membrane that is classified as a 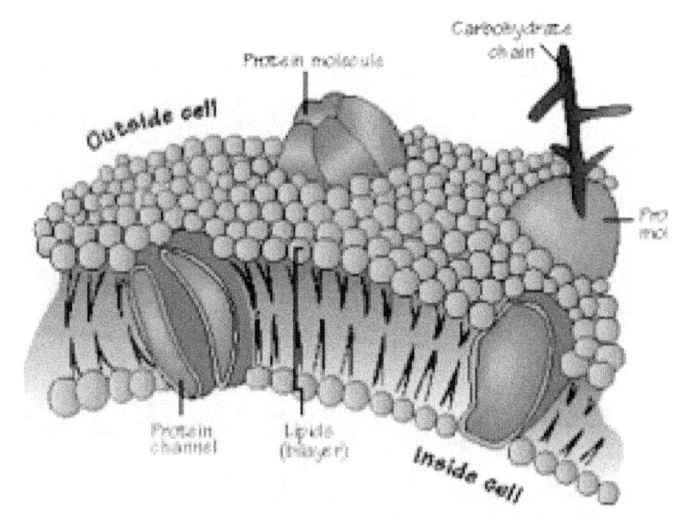 **phospholipid bi-layer** or a membrane with two layers of lipids. The fatty acid tails of phospholipids are **hydrophobic** or water-fearing. They point toward the inside of the bilayer. The phosphate molecules are hydrophilic or water loving and face outward. They come in contact with water and fluid that surround the cell.

The cell membrane is a semi-permeable membrane. **Semi-permeable** means that it only allows certain molecules to pass through it by diffusion. **Diffusion** is the movement of a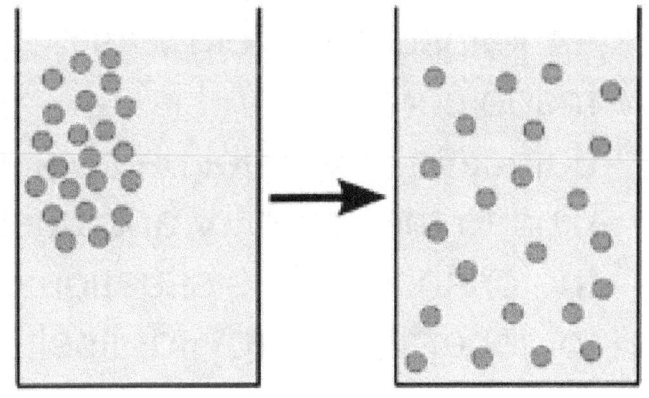

substance in the state of a liquid or a gas. Gas and liquids move from an area of high concentration to an area of low concentration. Small molecules such as oxygen, carbon dioxide and water are able to pass in-between the

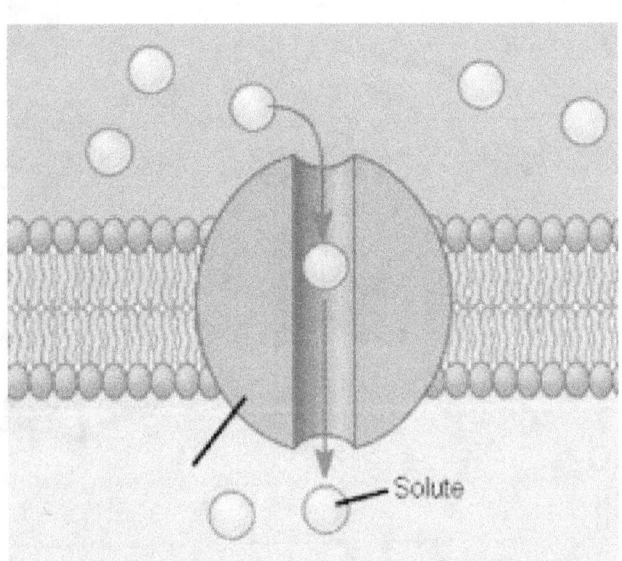

phospholipid molecules. Larger molecules such as glucose, a sugar, are allowed into the cell via **protein channels** that are embedded into the phospholipid bi-layer. Protein channels allow larger molecules to pass into and out of the cell and require energy to open and close them. An example of this is the glucose channel that exists in the cell membrane.

Insulin, a special protein, binds to a receptor on the surface of the cell, opens and closes this channel to allow glucose to pass into the cell. If the insulin protein was not made correctly and is defective, it will not be able to bind to the receptor and open the channel.

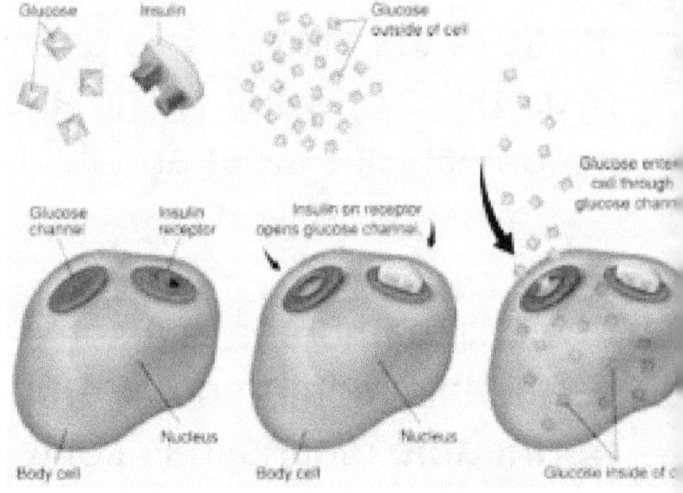

This situation causes the cell not to be able to generate energy for itself by being broken down in the mitochondria. Instead, the glucose accumulates in the blood and raises the blood sugar.

Knowledge and Comprehension

1. What is the function of the cell membrane?

2. What type of molecules make up the cell membrane?

3. Explain how the phosphate molecules and fatty acids in phospholipids arrange themselves to create the cell membrane.

4. Explain why the cell membrane is semi-permeable.

5. Explain what protein channels in the cell membrane do.

6. Explain what occurs if insulin molecules are deformed and cannot bind to glucose channels on the cells.

Types of Cells

A **cell** is the smallest functional and structural building block of life. Cells make up all living things. They communicate with each other and work together to make up the structure of living things and to carry out the many functions of organism.

There are two different types of cells. Cells that have a nucleus and cells that do not have a nucleus. The **nucleus** is a structure within the cell that stores all the genetic information known as **DNA** for the function and reproduction of the cell.

Eukaryotic cells have a nucleus. They have genetic material DNA. Plant cells and animal cells are examples of eukaryotic cells.

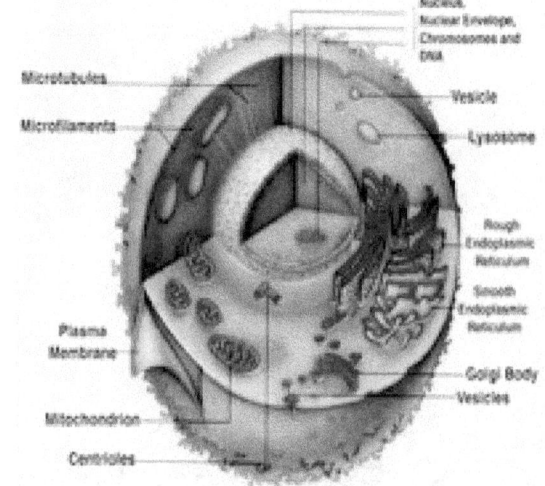

Prokaryotic cells or prokaryote do not have a nucleus. Prokaryotic cells are classified into two groups: bacteria and archaea.

It is believed by scientists that the

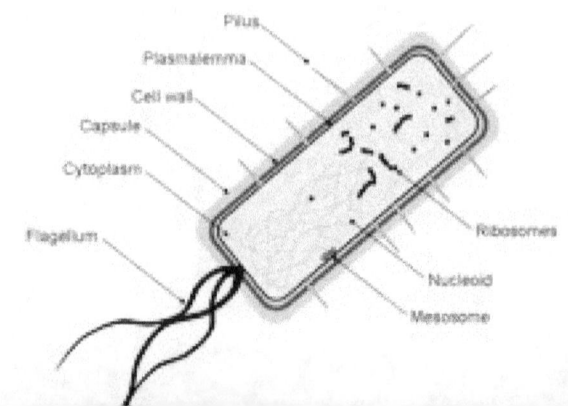

first living things on Earth were probably prokaryotes. More complex life forms are said to have evolved from these single celled organisms. The evidence supports this fact is that microbials, 3.48 billion years old, were discovered in western Australia. Other evidence that supports this is the research that has been done by biochemist Douglas Theobaldis who calculated that the last universal common ancestor of all life forms on Earth is lease 10^{2860} more probable than having multiple ancestors. This evidence suggests that life on Earth followed an evolutionary pattern suggested by Charles Darwin in his book **On the Origin of Species**. In his book, he infers that all living organisms on Earth arose through evolutionary processes from one primordial form.

Sources:

http://news.nationalgeographic.com/news/2010/05/100513-science-evolution-darwin-single-ancestor/

Knowledge and Comprehension
Words to Know:

Cell:

Nucleus:

DNA:

Eukaryotic Cells :

Prokaryotic cells:

1. What is a cell?

2. What are the two major types of cells?

Application, Analysis, Evaluation and Synthesis

3. Explain how eukaryotes and prokaryotes are different from each other. What do these organisms have in common?

4. Could a living organism survive without a nucleus? Find evidence from the text to support your answer.

5. Do you agree with the claim "All complex organisms evolved from a single-celled organism." Find evidence from the text to support your answer.

What is a Plant Cell?

Plants are **eukaryotes** or a multicellular organisms. **Plant cells** have unique features when compared to animal cells. They have most of the organelles and structures that exist within the animal cell, but also have a cell wall, chloroplasts, and a large central vacuole that have specialized functions to carryout the process of photosynthesis. **Photosynthesis** is the process of a plant making the simple sugar glucose with the light energy from the sun.

Plants have organelles called **chloroplasts** which is the location of photosynthesis. The chloroplasts use carbon dioxide and water molecules and convert them glucose through a series of chemical reactions. The chloroplasts contain **thylakoid membranes** which house the green pigment chlorophyll. **Chlorophyll** absorbs or takes in light energy within its molecular bonds. Electrons are then transferred from chlorophyll molecules to reaction centers

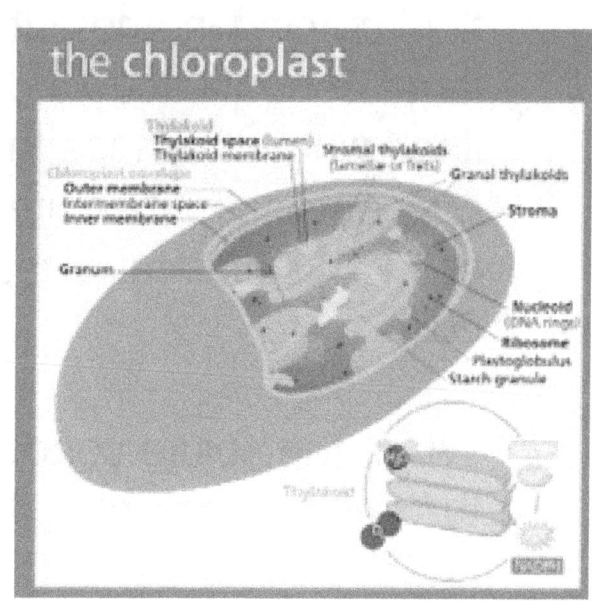

where it is used in photosynthesis.

The **cell wall** is a structure that surrounds the cell membrane of plant cells. The cell wall is rigid. It maintains the shape of the cell and protects the cell fro damage. Cell walls allow fluids and molecules to pass in and out of the cell. Unlike the animal cell, which is made of phospholipids and proteins, the cell wall is made mostly of a carbohydrate called **cellulose** and protein.

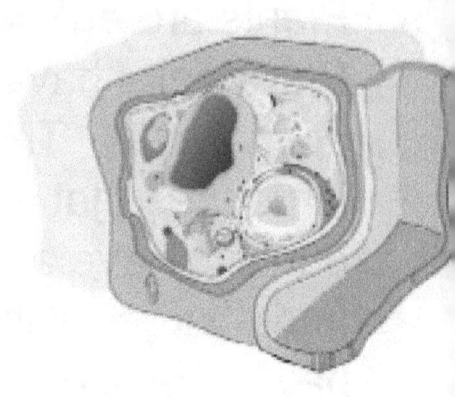

The **large central vacuole** maintains the internal pressure of fluids within the plant cell. This structure is a water-filled membrane that regulates the amount of fluid coming in and out of the cell. The large central vacuole draws water into itself when it senses that the concentration of solutes is low outside cell. This maintains the rigidity of the cell.

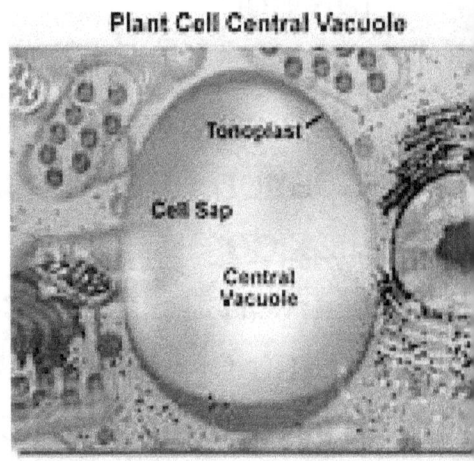

When the solute concentration is higher on the inside of the cell then the outside, the large central vacuole will release water out of the cell in order to decrease the concentration. In the plant cell, a solute is a mixture of

water and other substances such as molecules and minerals. These two actions are how the cells respond to changes in their environment in order to maintain equilibrium. **Equilibrium** is a state where the environments both inside and outside of the cells are in balance. In this case, the cells try to balance the solute concentration, and the pH of the solution through the movement of water in and out of the cells through a semi-permeable membrane. This process is known as **osmosis**.

Knowledge and Comprehension
Words to Know:

Eukaryotes:

Photosynthesis:

Chloroplasts:

Chlorophyll:

Thylakoid membranes:

Cell Wall:

Large Central Vacuole:

Cellulose:

Equilibrium:

Osmosis.

1. What three organelles or structures are unique to the plant cell when compared to an animal cell?

2. What is osmosis?

Application, Analysis, Evaluation and Synthesis

3. What function does the plant cell serve? How are the cell wall in a plant and the outer membrane of the animal cell different?

4. How does the plant cell maintain equilibrium?

5. If the solute concentration is higher in side the cell than it is outside the cell, what must occur for the cell to achieve equilibrium?

6. Predict what would occur if the central vacuole inside the plant cell was damaged? What would happen to the cell?

7. What is a chloroplast and what is its function within the plant cell? Could the cell function without the chloroplast? why or why not?

8. What is chlorophyll and what is its function in the plant cell? What is its role in the process of photosynthesis.

The Cell Membrane

The cell membrane is the protective barrier that surrounds the cell. It is a flexible covering that is made up of one or more layers of phospholipid molecules bonded or linked together. These layers are made of **phosholipids**. Lipids are fat. Phospholipids are made of a phosphate molecule, a glycerol molecule, and 2 fatty acid chains. The animal cell has a cell membrane that is classified as a 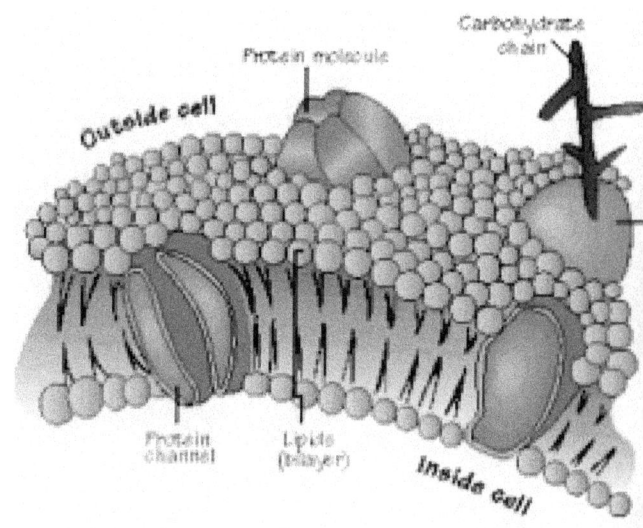 **phospholipid bi-layer** or a membrane with two layers of lipids. The fatty acid tails of phospholipids are **hydrophobic** or water-fearing. They point toward the inside of the bilayer. The phosphate molecules are hydrophilic or water loving and face outward. They come in contact with water and fluid that surround the cell.

The cell membrane is a semi-permeable membrane. **Semi-permeable** means that it only allows certain molecules to pass through it by diffusion. **Diffusion** is the movement of a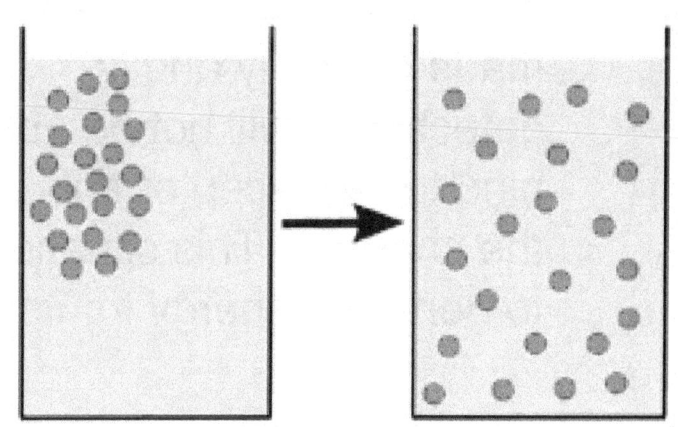

substance in the state of a liquid or a gas. Gas and liquids move from an area of high concentration to an area of low concentration. Small molecules such as oxygen, carbon dioxide and water are able to pass in-between the

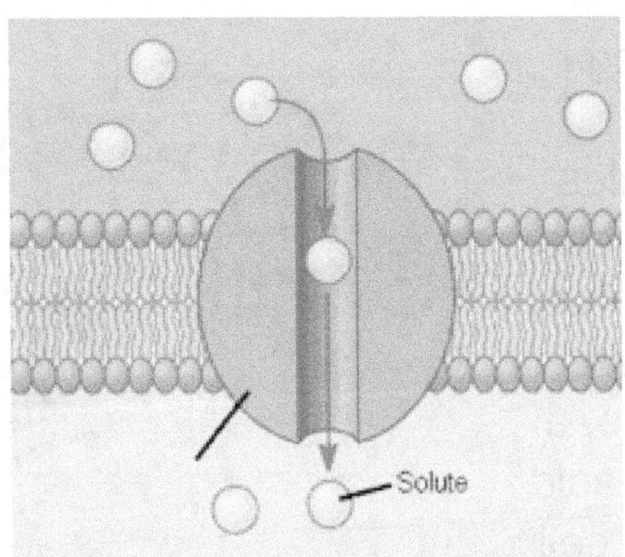

phospholipid molecules. Larger molecules such as glucose, a sugar, are allowed into the cell via **protein channels** that are embedded into the phospholipid bi-layer. Protein channels allow larger molecules to pass into and out of the cell and require energy to open and close them. An example of this is the glucose channel that exists in the cell membrane.

Insulin, a special protein, binds to a receptor on the surface of the cell, opens and closes this channel to allow glucose to pass into the cell. If the insulin protein was not made correctly and is defective, it will not be able to bind to the receptor and open the channel. This situation causes the cell not to be able to generate energy for itself by being broken down in the

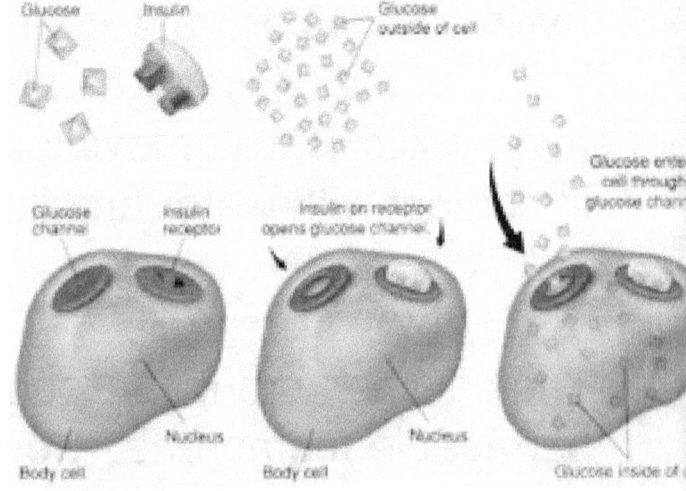

mitochondria. Instead, the glucose accumulates in the blood and raises the blood sugar.

Knowledge and Comprehension

1. What is the function of the cell membrane?

2. What type of molecules make up the cell membrane?

3. Explain how the phosphate molecules and fatty acids in phospholipids arrange themselves to create the cell membrane.

4. Explain why the cell membrane is semi-permeable.

5. Explain what protein channels in the cell membrane do.

6. Explain what occurs if insulin molecules are deformed and cannot bind to glucose channels on the cells.

The Nucleus of the Cell

The nucleus is a membrane-bound organelle that serves as the control center of the eukaryotic cell. A prokaryotic cells does not have a nucleus. It has a double membrane with **nuclear pores** that allow for the transport of larger molecules, such as RNA, in and out of the nucleus. Large molecules must be transported by carrier proteins. Small molecules, however, are free to move through the membrane. It houses genetic material known as **Deoxyribonucleic acid (DNA)** within it, maintains and repairs the DNA, and controls the activities of the cell through regulation of gene expression. DNA strands within the nucleus condense and wrap around proteins called **histones**. The wrapped DNA forms **chromosomes**.

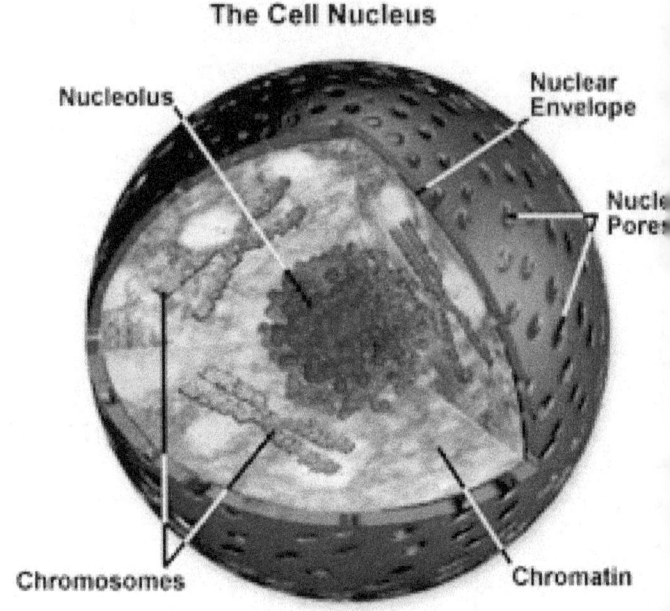

Figure 1

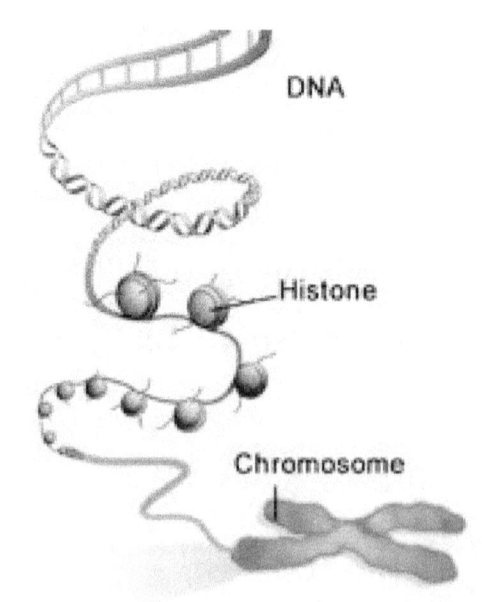

The nucleus controls the activities of the cell through gene expression. The genes within the chromosomes make up the genome or genetic instructions for the organism. **Genes** are specific segments of DNA that codes for proteins made by the cell. Genes are expressed by when they are turned on and off in response to chemical signals or special molecules and proteins, within the cell. The concentration or levels of these chemical signals will turn on or turn off the transcription of DNA within the gene. Transcription is the copying of the DNA to RNA molecules. RNA molecules are then **translated** to proteins within the ribosomes. Proteins regulate many activities and chemical reactions occurring inside the cell.

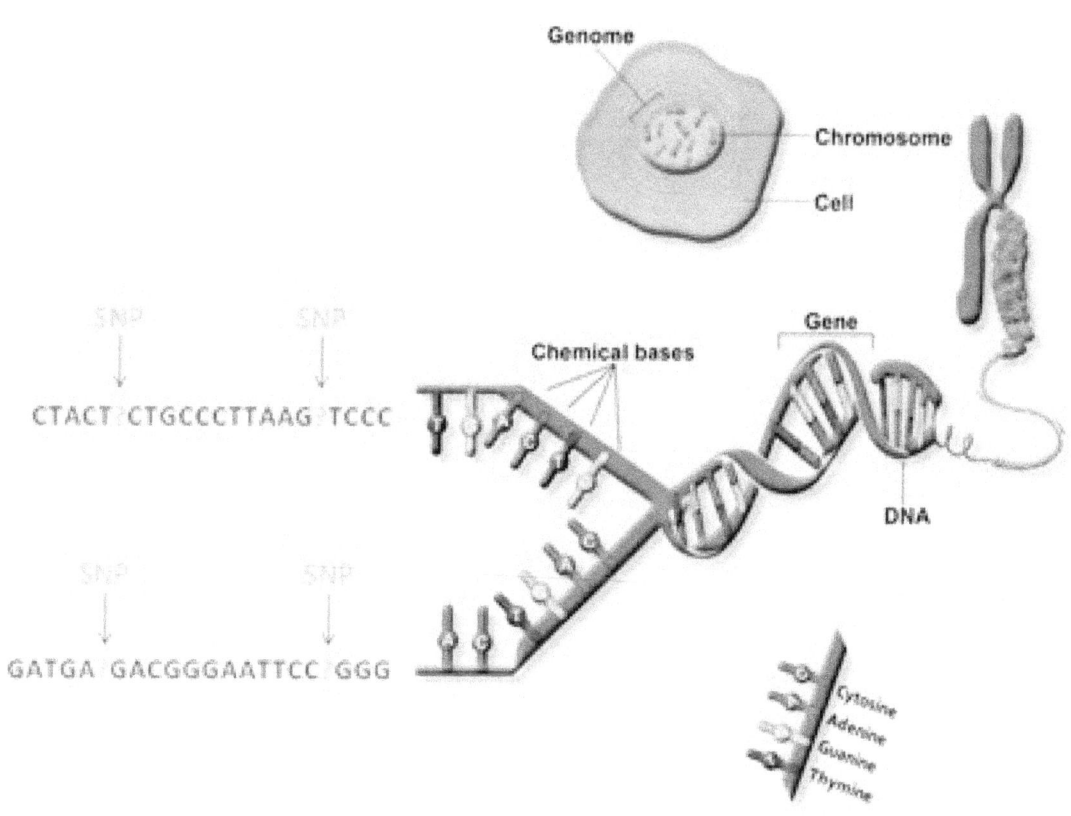

The nucleus also contains a number of proteins, RNA molecules, and a nucleolus. The nucleolus is made of protein and RNA molecules. It is the site of ribosome synthesis (making) and assembly (putting together). The **ribosomes** are protein complexes that **translate** or transform RNA molecules, that have been transcribed in the nucleus, to proteins. Ribosomes are transported into the cytoplasm of the cell after they are assembled by the nucleolus.

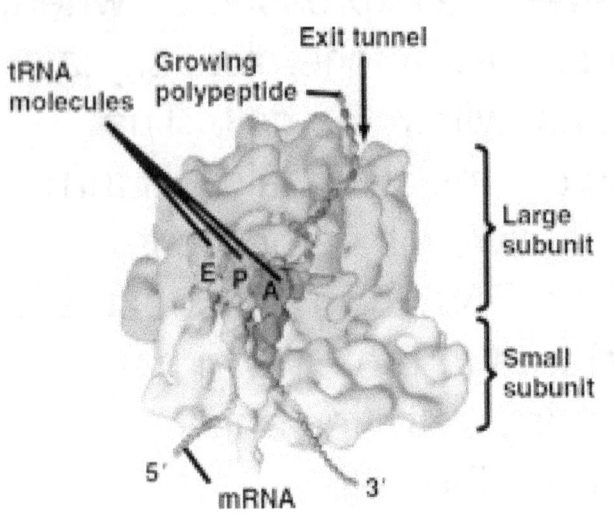

Knowledge and Comprehension:

1. What is the function of the nucleus?

2. Describe the structure of the nucleus.

3. What are genes?

4. Explain why the nucleus as known as a control center.

5. Explain how the expression of genes works.

6. What is the nucleolus? What is it's function?

The Mitochondrion

The **mitochondrion** is the powerhouse of the cell. It is an organelle that breaks down the sugar glucose to release stored energy through the biochemical process called cellular respiration. **Cellular respiration** is a set of biochemical reactions that converts the adenosine triphosphate (ATP) molecule into adenosine diphosphate (ADP). Energy that is stored within the bonds of the adenosine triphosphate (ATP) molecule is released during this reaction and used to drive the functions of the cell.

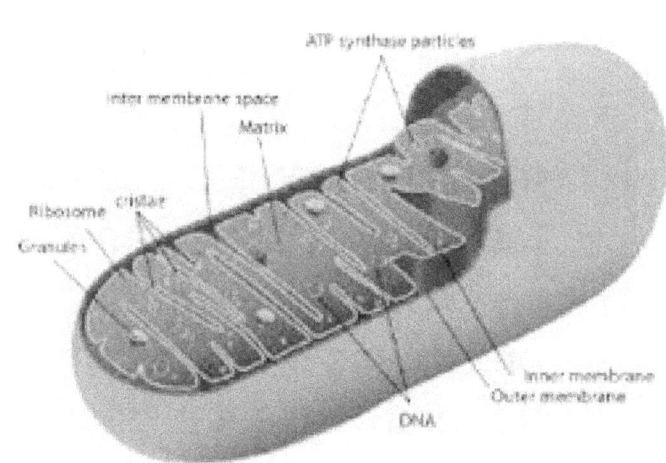

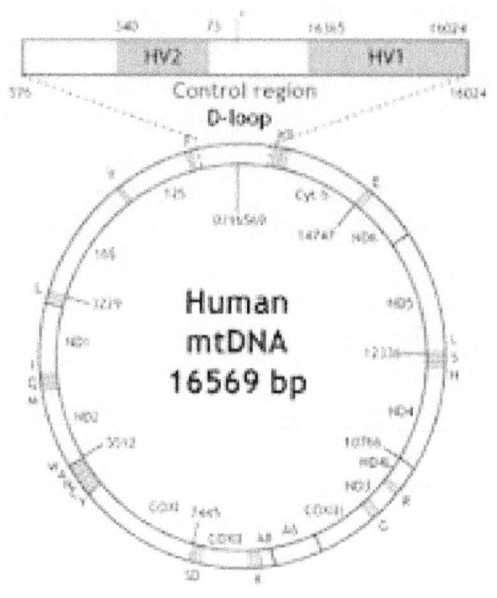

Scientists theorize that mitochondria may existed about 3.6 billion years ago as tiny archeobacteria or ancient bacteria. Evidence comes from the fact that these tiny organelles have their own genetic material, DNA, that exists in the form of a circular structure. Archeobacteria and bacteria both have their own DNA, some which exists in circular form. Mitochondria have their own ribosomes or protein making machinery. Through specific instructions from its DNA, it can manufacture its own proteins.

Scientists think that these bacteria, in an effort to protect themselves from the outside environment and ensure that they received a steady supply of glucose, allowed themselves to become engulfed by their prokaryotic host cell. Through the process called **endosymbiosis,** smaller organisms can live within their host cells without becoming harmful. By establishing a **symbiotic relationship**, both the cell and the mitochondria benefit each other and are able to live harmoniously with each other. The mitochondria break down glucose within the cell and prevent the inside of the cell from getting toxic. The cell, in return, receives a steady supply of energy from the breakdown of glucose and uses this energy to drive its chemical reactions and other cellular activities.

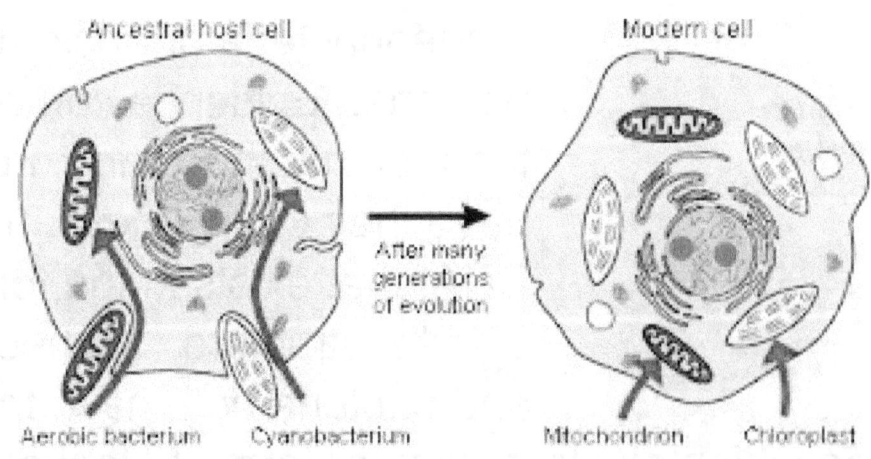

Knowledge and Comprehension
Words to Know:

1. What is the function of the mitochondrion ?

2. What is cellular respiration?

3. Explain how glucose is converted to energy in the mitochondrion. What is this energy used for?

4. What is the origin mitochondrion? What is the evidence for this?

5. Explain what was the relationship between the mitochondrion and the first cell.

The Endoplasmic Reticulum

The **endoplasmic reticulum** is an organelle that is made up of flattened sacs or tubes that are called **cisternae**. The endoplasmic reticulum is attached to the membrane of the nuclear membrane of the nucleus.

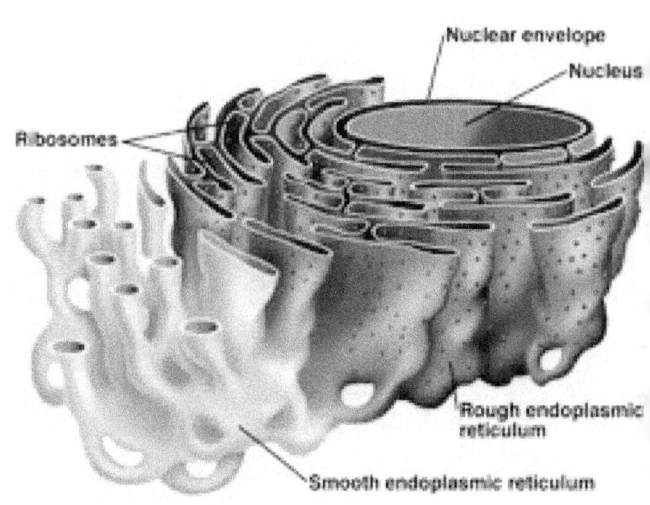

There are two types of endoplasmic reticulum: **smooth** and **rough**. **Smooth endoplasmic reticulum** does not have ribosomes. It has special functions within the cell such as metabolism or the breakdown of lipids (fats) or carbohydrates. It also has a special role in the detoxification of liver cells and the sex cells.

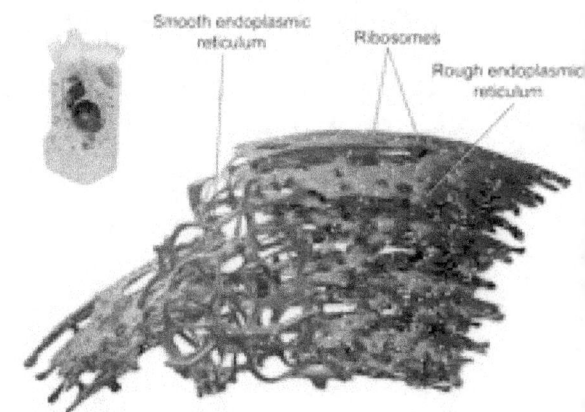

Endoplasmic Reticulum

Rough endoplasmic reticulum has ribosomes embedded into the membrane. **Ribosomes** are protein complexes that are the site of protein synthesis and assembly. Ribosomes attach amino acids together using an RNA template that is made from

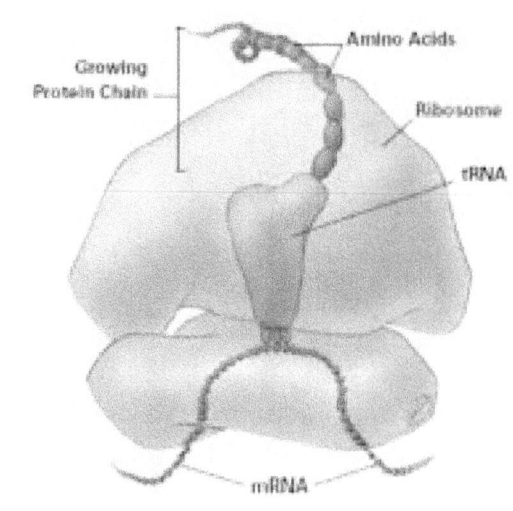

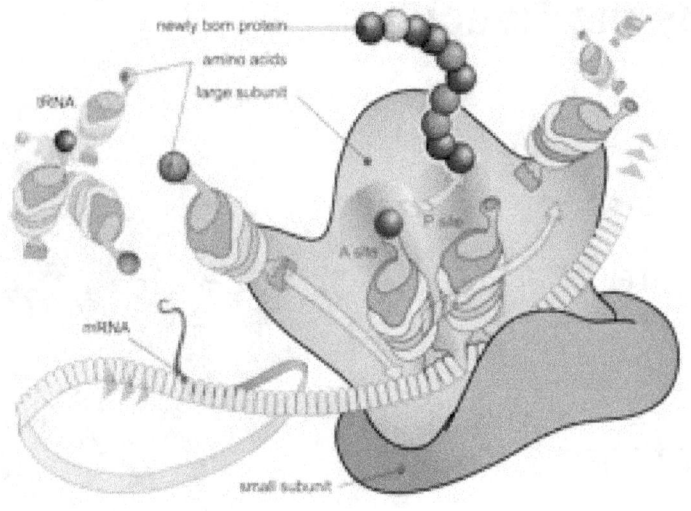

transcribing and copying a segment of DNA that encodes a protein. The protein is a gene product or a protein product that is encoded by a gene or a segment of DNA. The DNA is transformed through transcription into RNA in the nucleus. The RNA is then sent to the ribosome where it becomes the template for synthesizing or making the protein. Amino acids are collected by transfer RNA and set onto the template. The amino acids are then joined together by special enzymes or protein known as **RNA polymerases**.

Knowledge and Comprehension Questions

1. Where is the endoplasmic reticulum located in the cell?

2. What does the rough endoplasmic reticulum have that the smooth endoplasmic reticulum does not have?

3. What are the functions of the smooth endoplasmic reticulum?

4. What is the function of the ribosome?

5. Explain how chains of amino acids are made in the ribosome.

6. Explain why the endoplasmic reticulum is located next to the nucleus of the cell.

The Golgi Apparatus

The **Golgi apparatus**, an organelle that present in most eukaryotic cells such as plant an animal cells, is part of an endomembrane system. The membrane is located in the cytoplasm or the fluid inside the cell. It is made up of stacks that are membrane bound structures known as **cisternae**. Every cisternae is a flat, membrane bound disc that includes special proteins called enzymes which help modify or change proteins from the endoplasmic reticulum.

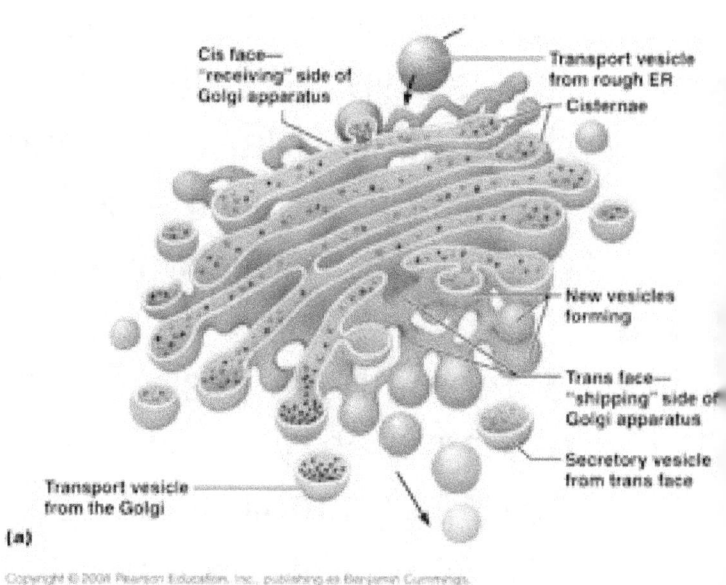

The function of the Golgi apparatus is to process proteins and other large molecules inside the cell. Proteins that were made within the ribosomes inside the endoplasmic reticulum. It also packages these proteins for transport to their final destination.

Vesicles containing proteins from the endoplasmic reticulum fuse to the **cis face** of the golgi apparatus. The proteins the move through the cisternae to the part of the Golgi apparatus known as the **trans face**. It is in this part of the apparatus which the proteins are loaded into vesicles. The vesicles then pinch off from the membrane of the apparatus and is shipped to its final destination. Many of these vesicles move to the cell membrane and release their contents outside the cell.

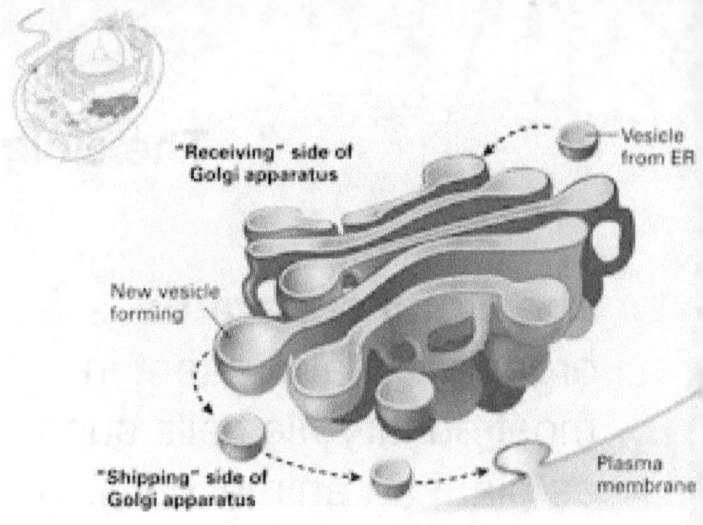

Some of these vesicles are lysosomes. They contain digestive enzymes within them. These enzymes break down large molecules such as carbohydrates and fats, damaged cells, and waste products from the cells.

Knowledge and Comprehension Questions

1. Identify where the Golgi apparatus located inside the cell.

2. Describe the function of the Golgi apparatus.

3. Explain how proteins from the endoplasmic reticulum enter the Golgi apparatus.

4. Explain how the Golgi apparatus modifies proteins sent from the endoplasmic reticulum.

5. Explain what lysosomes are and what they do.

6. Explain how waste products from the cell are broken down and transported out of the cell.

7. Explain how a vesicle forms and what its function is.

Cell Organelles and their Functions in Plant and Animal Cells

Nucleus: is a large organelle inside the cell where genetic material or DNA is stored and replicated (copied). A dark spot within the nucleus, called the nucleolus, is the location where ribosomes, organelles that make proteins, are made.

Ribosomes: organelles that make **proteins** from **amino acids** for the cell.

Endoplasmic Reticulum or ER: is a system of folded membranes in which proteins, lipids, and other materials. The endoplasmic reticulum transports and delivers substances to different places in the cell.

Mitochondria: is the powerhouse of the cell. It is an organelle that breaks down sugar to release energy. This energy is stored in the adenosine triphosphate (ATP) molecule.

Golgi Complex: is an organelle that packages and moves proteins to where they are needed both within and out of the cell.

Lysosome: are vesicles responsible for digestion that found in animal cells. They contain specialized proteins called enzymes that get rid of waste products, foreign invaders, and organelles that have been worn out or damaged.

Chloroplasts: are organelles in the plant cells which use light energy, carbon dioxide, and water to make its own food (**glucose**) through the process of **photosynthesis**.

Large Central Vacuole: an organelle that stores water and other materials.

Structure of the Cell Study Cards

Directions: Cut out the boxes with the keywords in them. Draw a picture under the keyword. Write the definition for the keyword on the back of the box..

Cell Organelles and their Functions

in Plant and Animal Cells

Nucleus: is a large organelle inside the cell where genetic material or DNA is stored and replicated (copied). A dark spot within the nucleus, called the nucleolus, is the location where ribosomes, organelles that make proteins, are made.

Ribosomes: organelles that make **proteins** from **amino acids** for the cell.

Endoplasmic Reticulum or ER: is a system of folded membranes in which proteins, lipids, and other materials. The endoplasmic reticulum transports and delivers substances to different places in the cell.

Mitochondria: is the powerhouse of the cell. It is an organelle that breaks down sugar to release energy. This energy is stored in the adenosine triphosphate (ATP) molecule.

The Cell: Unit exam

A. Identify the Organelle or Structure:

1. An organelle that breaks down glucose molecules within the cell and releases energy.

2. An organelle that contains the genetic information of the cell.

3. An organelle in the plant cell that uses sunlight, carbon dioxide and water to make glucose.

4. An organelle that contains enzymes that help breakdown foreign invaders, cell parts and wastes.

5. An organelle that modifies, stores, and transports molecules within the cell and outside the cell.

6. A protein complex that assembles amino acids into proteins.

7. An organelle that makes fats and lipids, breaks down substances that are toxic, and packages proteins for the golgi complex. This organelle may also contain ribosomes.

8. A structure that is rigid and gives support to the cell. It surrounds the cell membrane of a plant cell.

9. A structure that is a protective barrier and encloses the cell.

a structure that is web-like, network of fibers that provides structure to the cell, and aids in cell division and cell movement.

10. A thick fluid, made mostly of water, contained in a cell.

11. A small sack that surrounds molecules and transports them inside and outside the cell.

12. A compartment that stores water and other molecules within the cell.

B. Identify the following Keywords;

cell:

cell theory:

organelle:

DNA:

Eukaryote:

Prokaryote:

Cellulose:

Glucose:

Phospholipid:

Lipid:

Amino acid:

Protein:

Photosynthesis:

C. Draw an animal cell and label the following structures:

Large Central Vacuole:

Nucleus:

Vesicle:

Chloroplasts:

Ribosomes:

Cell Membrane:

Endoplasmic Reticulum or ER:

Mitochondria:

Cell Wall:

Cytoskeleton:

Golgi Complex:

Lysosome:

D. Draw a plant cell and label the following structures:

Large Central Vacuole:

Nucleus:

Vesicle:

Chloroplasts:

Ribosomes:

Cell Membrane:

Endoplasmic Reticulum or ER:

Mitochondria:

Cell Wall:

Cytoskeleton:

Golgi Complex:

Lysosome:

E. Explain the difference between:

1. A prokaryotic cell and a eukaryotic cell

2. A lysosome and a vesicle

3. The endoplasmic reticulum and the golgi complex

4. The mitochondrion and the chloroplast

5. The nucleus and the ribosome

Golgi Complex: is an organelle that packages and moves proteins to where they are needed both within and out of the cell.

Lysosome: are vesicles responsible for digestion that found in animal cells. They contain specialized proteins called enzymes that get rid of waste products, foreign invaders, and organelles that have been worn out or damaged.

Chloroplasts: are organelles in the plant cells which use light energy, carbon dioxide, and water to make its own food (**glucose**) through the process of **photosynthesis**.

Large Central Vacuole: an organelle that stores water and other materials.

Structure of the Cell Study Cards

Directions: Cut out the boxes with the keywords in them. Draw a picture under the keyword. Write the definition for the keyword on the back of the box..

Activities of the Cell

Common Core Workbook

Monica Sevilla

Contents

What is Photosynthesis? An Introduction
The Chloroplast and the Process of Photosynthesis
Cellular Respiration
What is ATP
Transcription
Translation
Mitosis and the Cell Cycle
Cell Division (The Cell Cycle) Study Cards
Meiosis: Producing Gametes
Cell Differentiation

What is Photosynthesis?
An Introduction

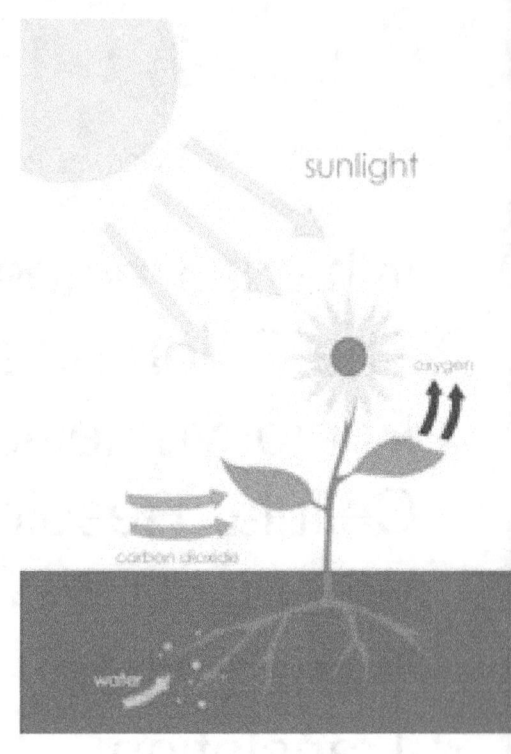

Photosynthesis is a process where photo or light energy is converted or changed by plants, cyanobacteria, and photoautotrophs such algae to **synthesize** or make glucose. An **autotroph** is an organism that makes its own food. A **photoautotroph** is an organism that makes its own food through photosynthesis. **Glucose** is a carbohydrate, a simple sugar, that is made of 6 carbon atoms that are arranged into a ring structure. This glucose is stored within plants and organisms and used as food.

Photosynthesis requires carbon dioxide (CO_2) from the atmosphere and water (H_2O) and light energy as the raw materials for this process. Sugars are produced through chemical reactions that are dependent on light as a source of energy. As the sugars are formed, oxygen gas is produced as a waste product. The following equation illustrates how many molecules of carbon dioxide and water are used to synthesize or make one molecule of glucose.

$$6CO_2 + 12H_2O + light \rightarrow C_6H_{12}O_6 + 6O_2 + 6H_2O$$

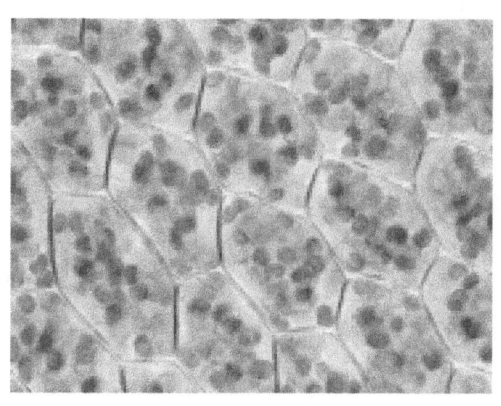

Scientists believe that the atmosphere of the early Earth became oxygen-rich through photosynthetic bacteria and algae living in the oceans millions of years ago. This transformation caused a major shift from a global **anaerobic** environment (no oxygen) to an **aerobic** (oxygen filled) environment, setting the stage for more complex animals and plants to evolve.

Knowledge and Comprehension
Words to Know:

Photosynthesis:

Synthesis:

Autotroph:

Photoautotroph:

Glucose:

Anaerobic:

Aerobic:

1. Describe what an autotroph is.

2. What is the difference between an autotroph and a photoautotroph.

Application, Analysis, Evaluation and Synthesis

3. What is photosynthesis. Summarize the process in your own words.

4. Which organisms are considered photoautotrophs? Do you think the process of photosynthesis is the same in each of these organisms? Why or why not?

5. How do scientists believe that oxygen was first formed on Earth? Do you agree with this theory? Why or why not?

The Chloroplast and the Process of Photosynthesis

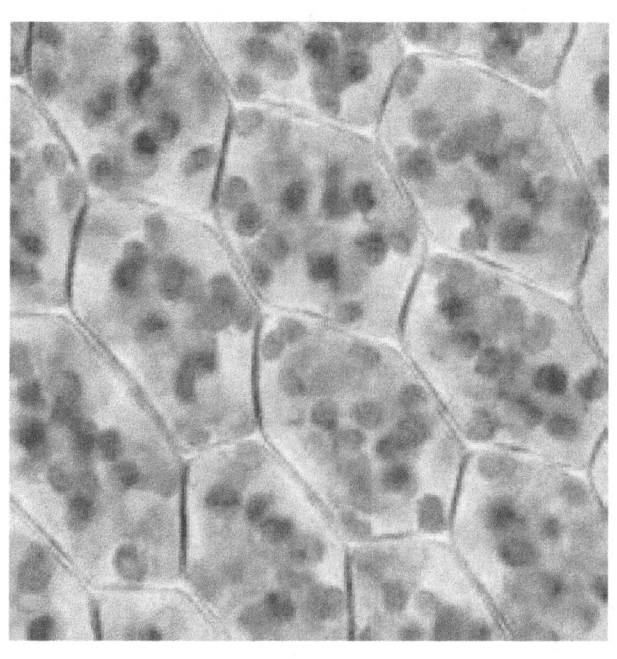

The plant cell contains important membrane-bound organelles within it. These organelles, known as **chloroplasts**, is the location of photosynthesis. Chloroplasts capture visible light energy from sunlight and convert it into sugars such as glucose that store energy. Sunlight, although it is white, contains many different wavelengths or colors of light. **Chlorophyll**, the pigment contained within the chloroplasts absorb or takes in some but not all the energy in visible light.

Chloroplasts have 2 different types of chlorophyll: chlorophyll a and chlorophyll b. These two pigments absorb red and blue wavelengths or colors of light. They absorb very little green color and is reflected. Plants are green because of the reflection of this color by chlorophyll.

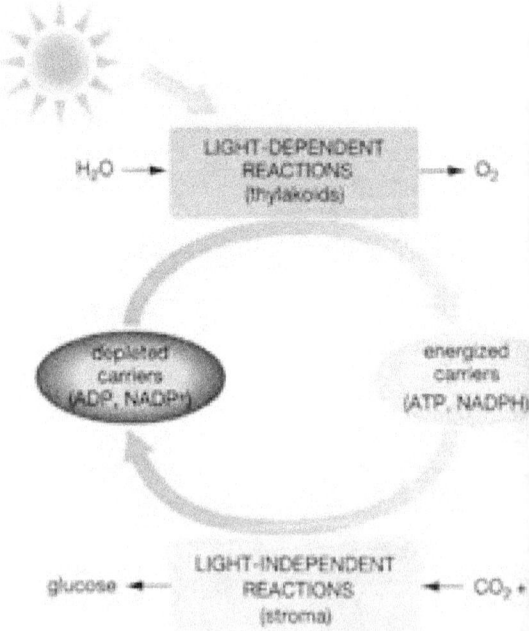

Anatomy of the Plant Cell Chloroplast

Figure 1

Most of the chloroplasts are located within the cells of the plant leaves. Their location within leaves maximizes the amount of area that is utililized by the chloroplasts to absorb light energy. The purpose of photosynthesis is to utilize energy from the sun to synthesize or make food in the form of sugar. this energy is crucial in order to drive and to power all the reactions associated with photosynthesis.

Two main parts of the chloroplast are needed for **photosynthesis** to occur: the grana and the thylakoid. The **grana** are stacks of thylakoid membranes. The **thylakoid membranes** are disc-shaped compartments that are enclosed by a membrane. Chlorophyll and protein is located within this membrane. Inside the chloroplast, the grana is surrounded by fluid called the **stroma**.

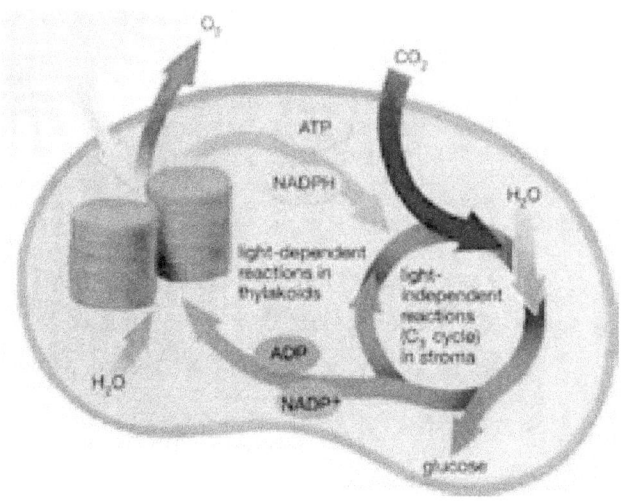

Two types of light reactions occur within the chloroplasts: light dependent reactions and light independent reactions.

Light dependent reactions capture the energy of the sun, breakdown water, and release oxygen. These reactions take place inside and across the thylakoid membranes. This energy is transferred to molecules such as ATP. ATP carries and stores this energy.

Light independent reactions use light energy from the light dependent reactions to make sugars. These reactions occur in the stroma. Carbon dioxide molecules are used to make two 3-carbon sugars and eventually, to build glucose, a 6-carbon molecule. Some of the energy that was captured from the sun is stored within this molecule and used as food for plants.

The combination of the light dependent reactions and light independent reactions together make up the process of photosynthesis. The overall equation for photosynthesis over-simplifies what occurs in this process. The equation

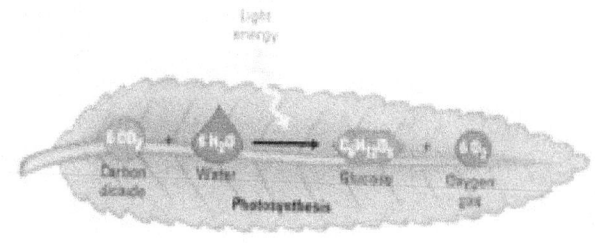

describes all the reactions that occur during the entire process.

Knowledge and Comprehension

Chloroplast:

Chlorophyll:

Grana:

Thylakoid Membranes:

Stroma:

Light Dependent Reactions:

Light Independent Reactions:

Photosynthesis:

Knowledge and Comprehension

1. For the following concepts, show how they are related to one another.

a. Chloroplasts and Photosynthesis

b. Chlorophyll and light energy

c. Grana and the Thylakoid Membranes

d. Light dependent reactions and Water molecules

e. Light independent reactions and Glucose

2. What is a chloroplast? What is its function? Where can it be found? Why is it important?

3. What important components exist within the chloroplast and why are they important to the process of photosynthesis.

Application, Analysis, Evaluation and Synthesis

4. Explain how light dependent and light independent reactions are different from one another.

5. Explain how light dependent and light independent reactions work together to create glucose molecules.

6. Explain why energy is important in the process of photosynthesis. Where does this energy come from?

Cellular Respiration

Cells, to have enough energy to carry out their functions properly, need to utilize energy on an ongoing basis. To provide this ongoing source of energy, cells use the process of cellular respiration. **Cellular respiration** is a set of chemical reactions that catabolize and convert macromolecules such as Amino acids, fatty acids, and the simple sugar **glucose** into useable energy to fuel the many activities of the cell. The activities that are powered by cellular respiration include the transport of molecules across the cell membranes, locomotion, and the synthesis or making of molecules and macromolecules in the cell.

The goal of cellular respiration is to **catabolize** or break down macromolecules and to release the energy stored within its high energy molecular bonds. During **glycolysis**, glucose molecules are broken down in the presence of oxygen, which acts as an **oxidizing agent** or an electron acceptor. Glucose molecules are converted into the high energy molecules **pyruvate** or CH_3COCOO^- and hydrogen ions (H^+).

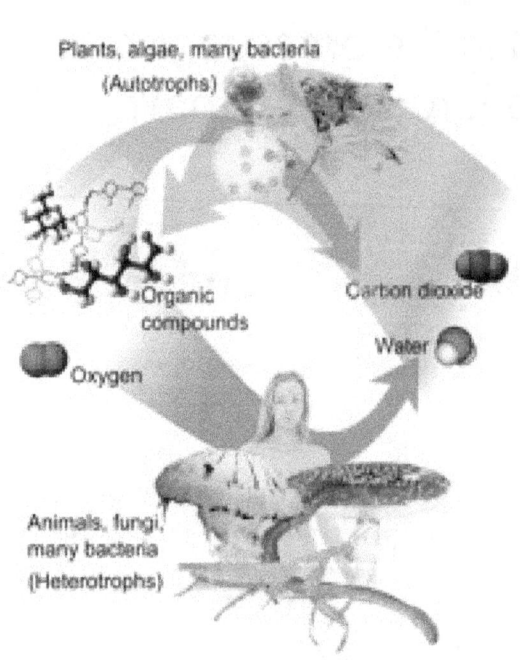

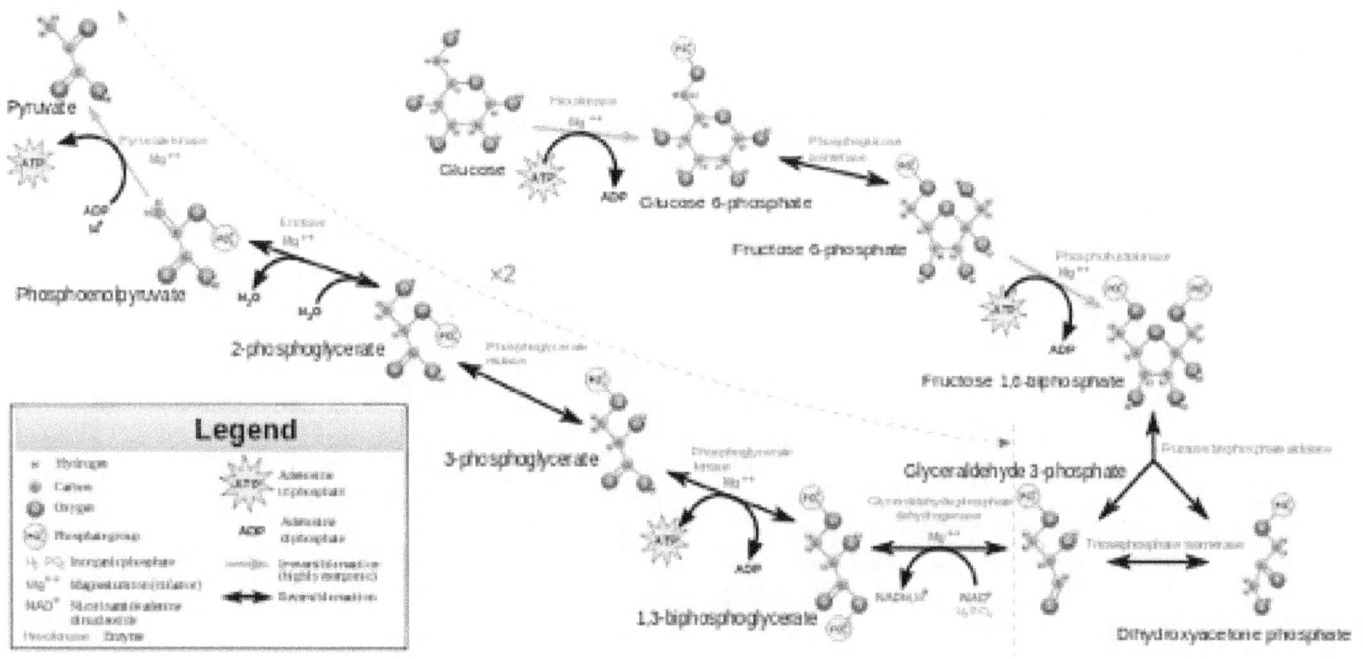

The energy that is released in this reaction is used to form **adenosine triphosphate (ATP)** and NDH. Glycolysis is a process that involves ten different reactions that are all driven by **enzymes** or special proteins that act as catalysts and lower the activation energy of chemical reactions. Glycolysis occurs within the **cytosol** or liquid portion of the cell.

Glycolysis occurs in nearly all organisms on earth. It is referred to by scientists as an ancient metabolic pathway that was used by organisms that existed within the ocean during the Archean eon 2.5 billion years ago. The environment at this time was hot, volcanic, and lacked oxygen. The process of glycolysis during this time was catalyzed by metals and in the presence of enzymes within the aqueous environment of the ocean.

Source:
http://en.wikipedia.org/wiki/Cellular_respiration

Knowledge and Comprehension

Cellular Respiration:

Glucose:

Catabolize:

Glycolysis :

Oxidizing Agent:

Adenosine Triphosphate (ATP):

Pyruvate:

Enzyme:

Cytosol:

1. What is cellular respiration?

2. What is glycolysis?

3. What element must be present in order for glucose to be broken down into pyruvate?

Application, Analysis, Evaluation and Synthesis

4. What is the role of ATP in glycolysis? Explain why it is important to the cell.

5. Explain what an enzyme is and why they are important.

6. Explain how the process of glycolysis first developed on Earth.

What is ATP?

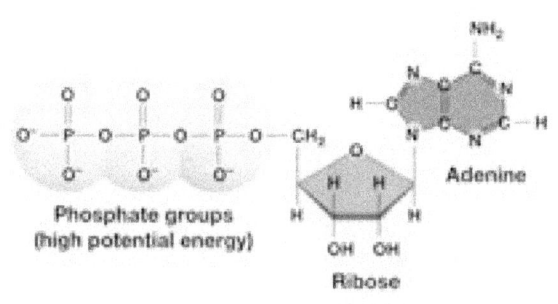

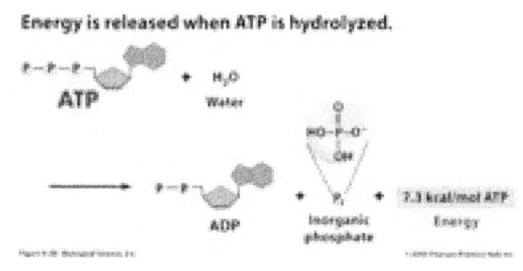

Every cell uses energy that is carried in the molecule of ATP. **Adenosine Triphosphate**, or **ATP**, contains chemical energy within its bonds. This energy can only be used when ATP is **metabolized** or broken down. The energy that is released is used to power the activities of the cell. Some of these activities include the active transport of molecules through the cell membrane and building molecules such as proteins, hormones, and fatty acids to help build more cells and regulate the many chemical reactions that occur within the cell.

Every time an atom creates a **bond** by sharing its electrons with another atom, energy is used. **Electrons** are the sub-atomic particles that exist around the outside of an atom's nucleus. This energy is stored within the bond that holds the shared electrons together. This occurs

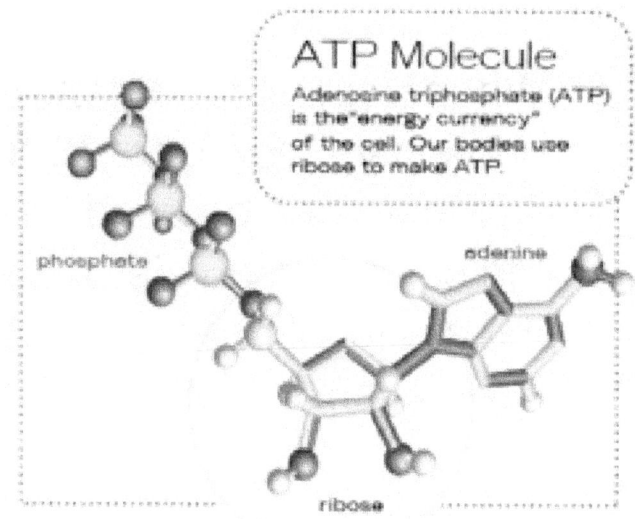

The energy molecule ATP is a ribose molecule joined to two other molecules. No ribose, no ATP. No ATP, no energy.

because electrons both have a negative charge and repel each other electrically. To keep them close enough together in the same physical location, energy is used to counteract the force of repulsion.

ATP is made up of an adenosine molecule and 3 phosphate groups (-P) attached to it. The 3 phosphate groups are attached together in a chain. When the bond between one of the phosphate groups in the chain is broken, **energy** or the ability to perform work, is released and the cell is able to power its activities.

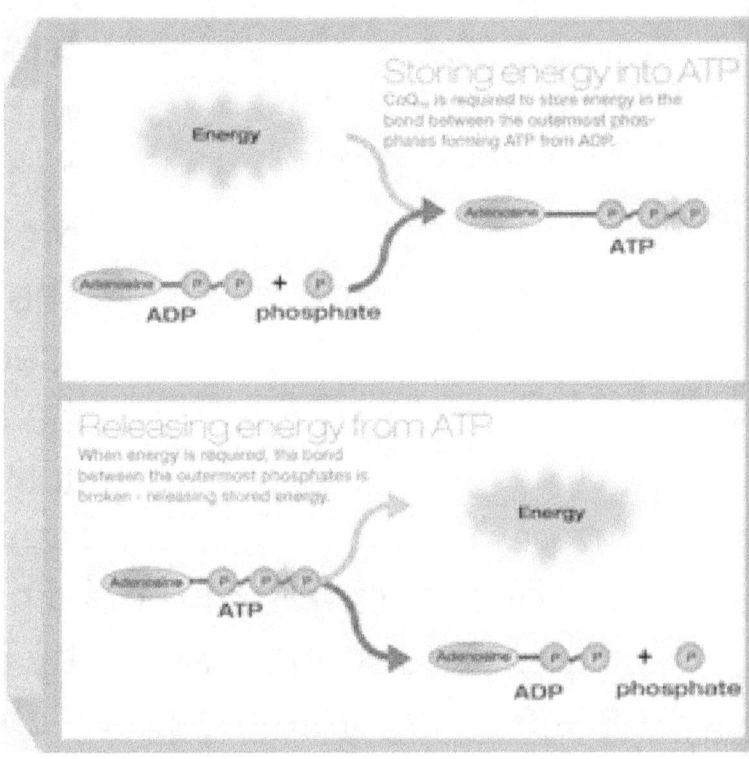

ATP can also be made, stored, and used later

When a phosphate group attaches to ADP, adenosine di-phosphate, energy is absorbed and ATP is formed. The ATP can then be used for energy in the future. In this way, the cell can regulate or control the energy it makes and the energy it uses to maintain **homeostasis** or a stable internal environment.

Source:
http://en.wikipedia.org/wiki/Cellular_respiration

Knowledge and Comprehension

Adenosine Triphosphate:

Metabolize:

Bond:

Electrons:

Energy:

Homeostasis:

1. What is ATP?

2. Where is chemical energy stored in the ATP molecule?

3. What is the energy released from ATP used for.

Application, Analysis, Evaluation and Synthesis

4. Explain how bonds between atoms are formed. What do bonds store?

5. What is the difference between ADP and ATP.

6. Explain how ADP is transformed into ATP.

Transcription

Transcription is the process in which DNA within the nucleus of the cell is copied into messenger RNA (m-RNA) used by the ribosomes to produce proteins. DNA is used as a template to produce a different nucleic acid that is complementary to DNA. This occurs through the action of the enzyme RNA polymerase. An **enzyme** is a special protein that carries out a specific function or job. In the case of RNA polymerase, **messenger RNA (m-RNA)** is produced by **polymerizing** or attaching **ribonucleotides** or RNA bases, together into a chain. The ribonucleotides include adenine, guanine, cytosine, and uracil. Uracil is used instead of thymine. Polymerization of the RNA begins at the 3' end of the RNA chain and proceeds toward the 5' end.

Control of transcription affects the expression of genes. Genes can be turned "on" or "off" by controlling the activity of DNA polymerase. Transcription is initiated at locations on the DNA called **promoters**. If an **inhibitor** is preset and binds to DNA polymerase, the shape of the enzyme changes and it falls off the DNA strand. Replication is terminated at locations on the DNA called **terminators**. The expression of the gene is turned "off" or inhibited and its gene product, a protein, is not produced. This mechanism is used as part of a communication and feedback system where the cell can respond to different conditions within the environment, and be able to adapt to the environment by producing more or producing less proteins.

Knowledge and Comprehension:

Transcription:

Enzyme:

Messenger RNA (m-RNA):

Polymerize:

Ribonucleotides:

Promoter:

Inhibitor:

Terminator:

1. Describe what transcription is.

2. Indicate what the function of an enzyme is.

3. Describe what occurs when a ribonucleotide or RNA base is polymerized.

Application, Analysis, Evaluation, and Synthesis

4. Explain why RNA polymerize is important.

5. Explain how gene expression is controlled through the process of transcription.

Translation

Translation is the process in which RNA is transformed into an amino acid chain or a **polypeptide chain**. Translation occurs through the action of the ribosome.

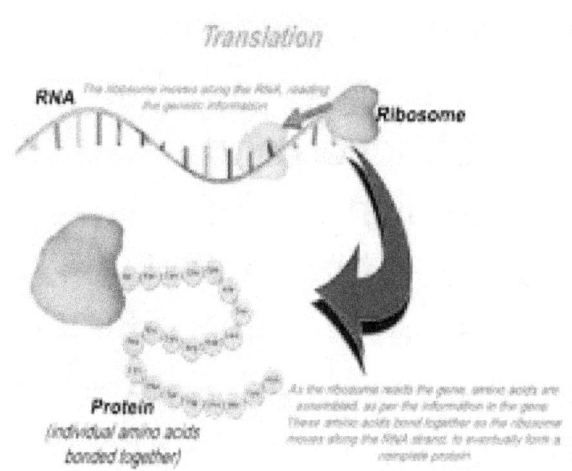

The **ribosome** is a large complex known as a ribonucleoprotein which contains two functional protein subunits: a ribosomal subunit that reads the mRNA sequence and a large subunit that links **amino acids**, the basic building blocks of proteins, together into a polypeptide chain. These two subunits work together as one unit, a complex protein complex. The ribosome is located within the cytosol or fluid part of the cell.

Ribosomes start the translation process by binding to **messenger RNA (m-RNA)**. m-RNA serves as a template in which amino acids are placed. The amino acids needed to build the polypeptide chain are selected and carried to the

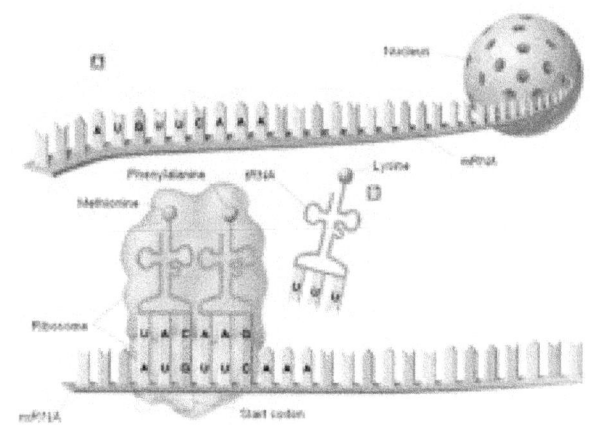

ribosome by **transfer RNA (t-RNA)**. The t-RNA binds directly to the m-RNA chain. Each amino acid is represented by a specific sequence of 3 ribonucleotides or RNA bases. Peptide bonds are formed between each amino acid within the ribosome creating a polypeptide or many peptide chain. The polypeptide is terminated when the ribosome encounters a stop codon. This codon induces the binding of a protein know as a release factor. The **release factor** promotes the destruction and disassembly of the ribosome. The resulting polypeptide chain is released and ready for transport by the vesicles. folded into a functional protein by special proteins. mRNA is used as a template for the assembly of an amino acid or polypeptide chain. RNA is complementary to DNA and is encoded by the DNA within the genes.

Knowledge and Comprehension:

Translation:

Polypeptide Chain:

Ribosome:

Amino Acids:

Messenger RNA (m-RNA):

Transfer RNA (m-RNA):

Release Factor:

1. What is translation?

2. Why is transfer RNA (t-RNA) needed for the assembly of amino acid chains?

Application, Analysis, Evaluation, and Synthesis

3. What is a ribosome? What is its function? Why is it referred to as a protein complex?

4. Describe the process of translation. How is an amino acid chain assembled in the ribosome?

5. What is the function of m-RNA? What is the function of t-RNA? How do they work together to make an amino acid chain?

Mitosis and the Cell Cycle

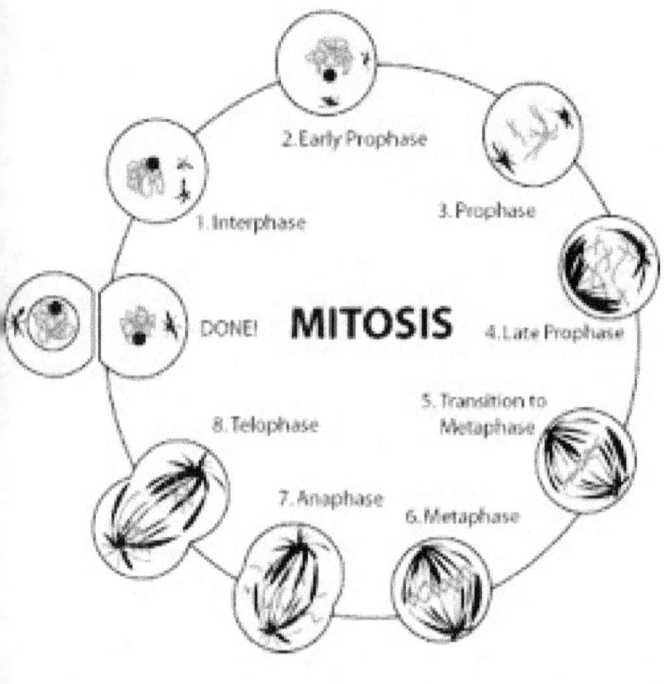

The life cycle of a cell is known as the **cell cycle**. Damaged cells are replaced with healthy, new cells within the organism on a constant basis. The cycle begins when the cell is first formed and ends when one cell divide to form new cells. The cell cycle includes distinct phases or steps that characterize how a cell replicates its genetic material or DNA and divides into two new cells. During the phase known as **interphase**, DNA is replicated or copied. **Mitosis,** or cell division, occurs as the 2 sets of DNA are moved to opposite poles within the cell. The two new cells that are formed are called daughter cells and are formed. The phases or steps of mitosis, in detail, include:

Interphase   Chromosomes are copied. Each chromosome becomes two chromatids.

Prophase   Mitosis begins. Chromosomes condense from long strands of DNA into rod-like structures.

Metaphase The nuclear membrane is dissolved. Paired chromatids align at the cell equator.

Anaphase The paired chromatids separate and move to opposite sides of the cell.

Telophase A nuclear membrane forms around each set of chromosomes, and the chromosomes condense. Mitosis is complete.

Cytokinesis The cell pinches off into separate cells. In a cell with a cell wall, a cell plate forms and separates the cell into two new cells.

Mitosis

Interphase
The nucleolus and the nuclear envelope are distinct and the chromosomes are in the form of threadlike chromatin.

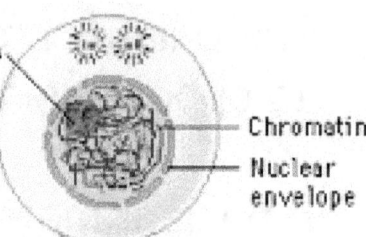

Prophase
The chromosomes appear condensed, and the nuclear envelope is not apparent.

Metaphase
Thick, coiled chromosomes, each with two chromatids, are lined up on the metaphase plate.

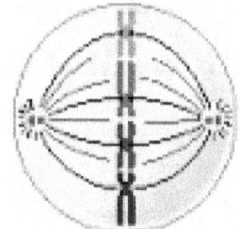

Anaphase
The chromatids of each chromosome have separated and are moving toward the poles.

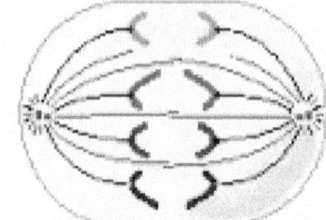

Telophase
The chromosomes are at the poles, and are becoming more diffuse. The nuclear envelope is reforming. The cytoplasm may be dividing.

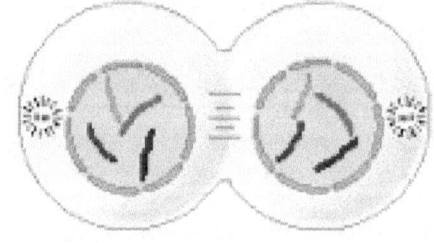

Cytokinesis
Division into two daughter cells is completed.

 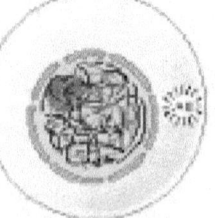

Knowledge and Comprehension Questions

1. What is the cell cycle? What is mitosis?

2. Explain why the cell cycle exists.

3. For each phase of mitosis, include the keyword, a picture, and a description of each of the phases.

Interphase

Prophase

Metaphase

Anaphase

Telophase

Cytokinesis

4. Why must DNA be replicated before mitosis occurs?

5. What would happen if the cell did not replicate its DNA and mitosis occurred?

Cell Division (The Cell Cycle)

The life cycle of a cell is known as the cell cycle. It begins when the cell is first formed and ends when cells divide to form new cells. During interphase, DNA is copied in the cell. **Mitosis** occurs as the the 2 sets of DNA are moved to opposite poles within the cell. Two new cells are formed as the center of the cell splits.

Interphase Chromosomes are copied. Each chromosome becomes two chromatids.

Prophase Mitosis begins. Chromosomes condense from long strands of DNA into rod-like structures.

Metaphase The nuclear membrane is dissolved. Paired chromatids align at the cell equator.

Anaphase The paired chromatids separate and move to opposite sides of the cell.

Telophase A nuclear membrane forms around each set of chromosomes, and the chromosomes condense. Mitosis is complete.

Cytokinesis The cell pinches off into separate cells. In a cell with a cell wall, a cell plate forms and separates the cell into two new cells.

Interphase
The nucleolus and the nuclear envelope are distinct and the chromosomes are in the form of threadlike chromatin.

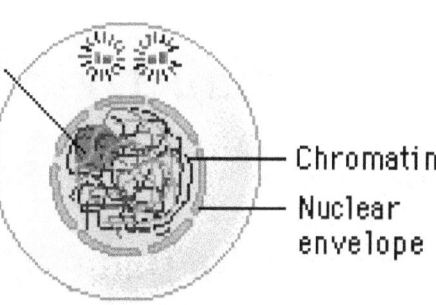

Prophase
The chromosomes appear condensed, and the nuclear envelope is not apparent.

Metaphase
Thick, coiled chromosomes, each with two chromatids, are lined up on the metaphase plate.

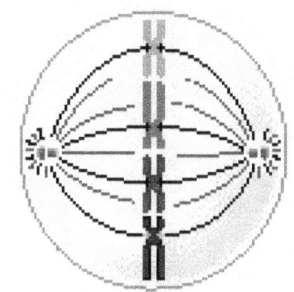

Anaphase
The chromatids of each chromosome have separated and are moving toward the poles.

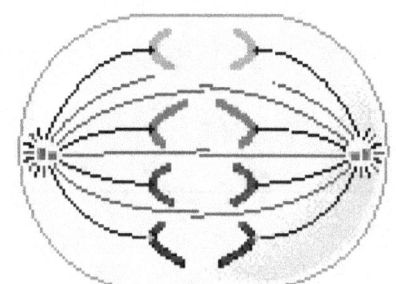

Telophase
The chromosomes are at the poles, and are becoming more diffuse. The nuclear envelope is reforming. The cytoplasm may be dividing.

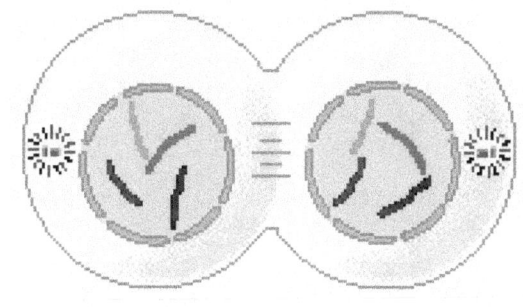

Cytokinesis
Division into two daughter cells is completed.

 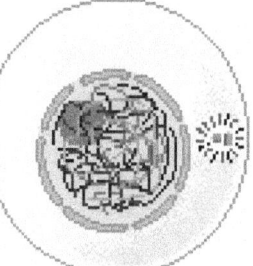

Instructions for Study Cards

For each phases of mitosis, make study cards that include the keyword, a picture, and a description of each of the phases.

Interphase

Prophase

Metaphase

Anaphase

Telophase

Cytokinesis

Meiosis: Producing Gametes

Meiosis is a special type of cell division that produces gametes or sex cells for the purpose of sexual reproduction. **Sexual reproduction** allows the genomes of male and female organisms to combine and create new individuals that genetically represent both their parents. Meiosis reduces the number of chromosomes in the cells, by one half. Humans have 2 sets of 24 chromosomes or these cells are referred to as being **diploid**. The process of meiosis effectively reduces the number of the chromosomes this to 1 set or 24 chromosomes. The cells are called **haploid**. When the sperm fertilizes the egg, the DNA within the two gametes, or sex cells, combines together and all the new cells that are produced through **cell division** will make up the developing embryo. All these new cells will **differentiate** or change into all the somatic cells and tissues of the body.

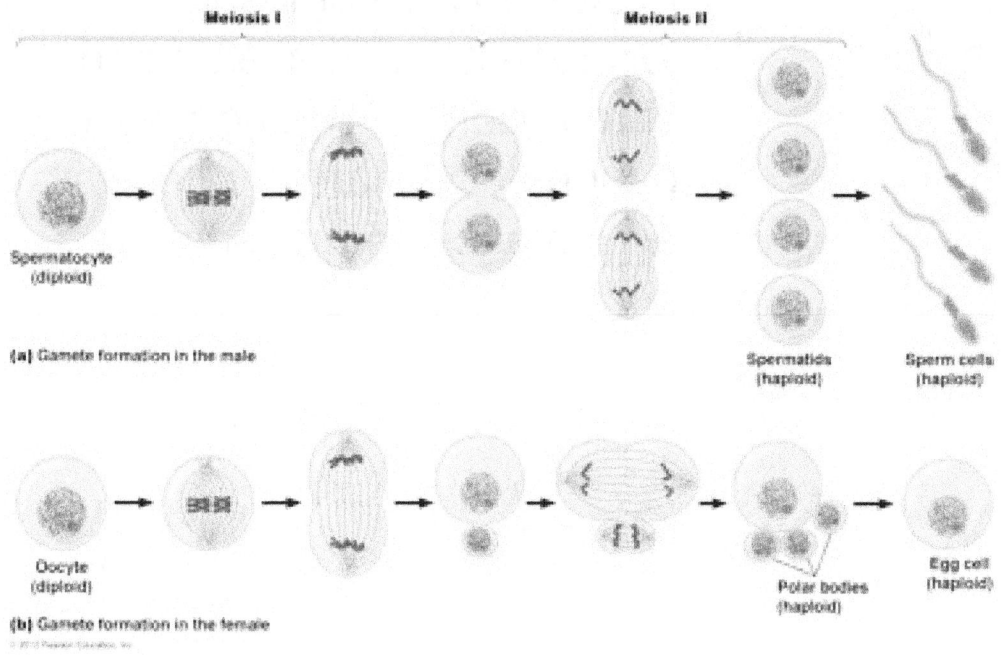

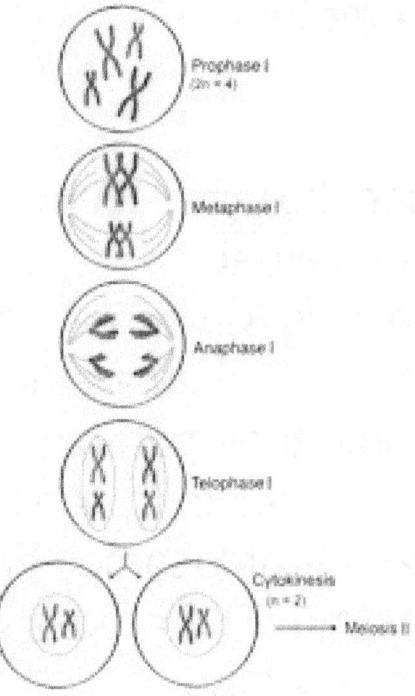

During the first phase of meiosis, recombination of DNA can occur. **Recombination** or the recombining of DNA of homologous or similar chromosomes during meiosis allows crossover to occur. **Crossing over** is the process of one chromosome exchanging its DNA with another chromosome. Through cell division, daughter cells are created with 24 recombined chromosomes.

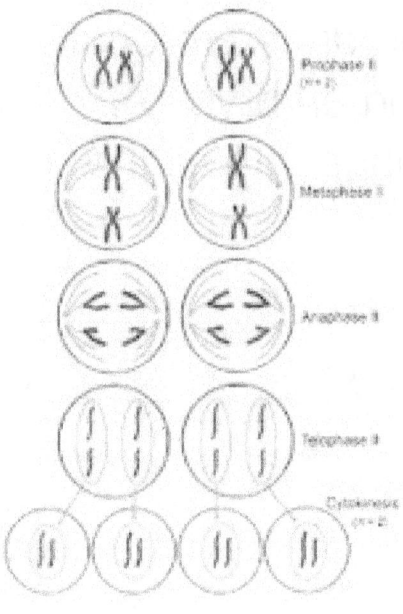

During the second phase of meiosis, the recombined chromosomes are broken apart into chromatids. A **chromatid** is one of the arms that makes up a chromosome. 2 arms join together to make up one chromosome. Through cell division, daughter cells are created with one arm of each chromosome.

Sources:

http://en.wikipedia.org/wiki/Meiosis

Knowledge and Comprehension:

Meiosis:

Sexual Reproduction:

Fertilization:

Egg:

Sperm:

Gametes:

Recombination:

1. Identify the goal of meiosis.

2. Describe the process of meiosis. How does it work?

3. Identify what a gamete is. What are the two gametes that are used in sexual reproduction?

4. What is the difference between a haploid and a diploid cell? Why is this important in sexual reproduction.

5. What is the difference between the two phases of meiosis? How does each achieve the goal of producing gametes?

The Differentiation of Cells

Cell differentiation is the process by which a cell changes into a specialized cell in multi-celled organisms. Differentiation first occurs when an organism develops from a **zygote**, a developing animal, into a more complex form, also in adulthood. **Embryonic stem cells**, cells that have not transitioned into different cell types, undergo a process by which certain genes are turned on and others are turned off transforming them into different cell types over a predetermined period of time. Each of these cell types have a different purpose and carryout different functions. These cell types eventually lead to the different tissues, organs, and organ systems within an animal.

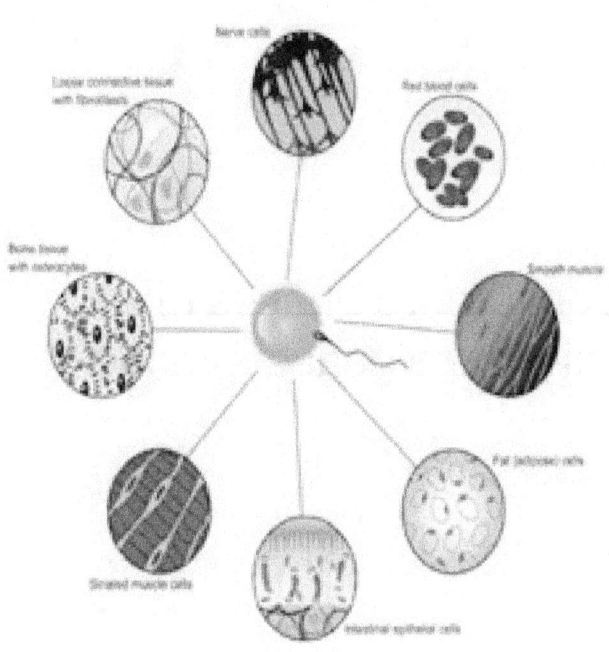

In order for cell differentiation to occur, the cells become different in size, metabolic activity, shape and membrane potential. These changes are the result of the expression of certain genes within the DNA of each cell. **Gene expression** is the process by which information is used to

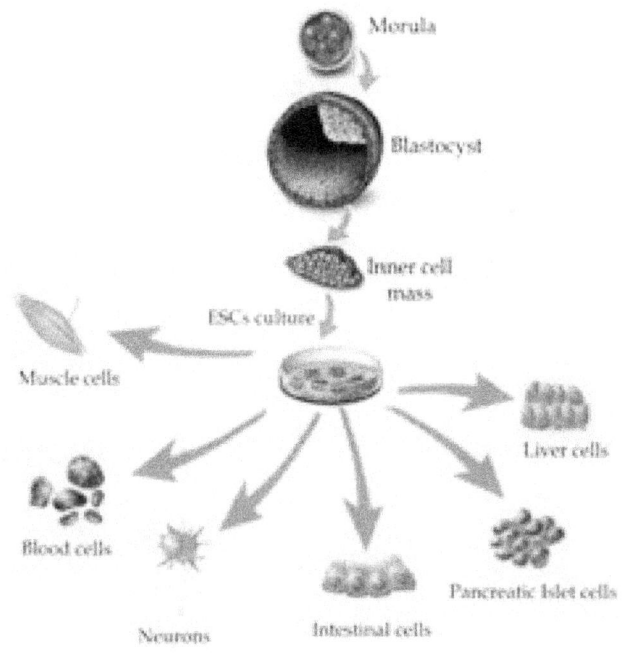

make a product that has a defined **function** or job such as a protein or an enzyme.

Scientists have been able to take a sample of stem cells from the early developmental stages of a human embryo and grow them in culture.

Embryonic stem cells were take from inner cell of the blastocyst stage and grown in the presence of growth factors in petri dishes. These cells eventually differentiated into the cell types that represent the muscles, nervous system, and the circulatory system, and organ system. Specific cell types that were observed included muscle cells, blood cells, neurons, intestinal cells, pancreatic islet cells, and liver cells.

As the stem cells grow and mature, a set of successive changes in the expression of its genes occurs. The genes code for bio-chemicals called **growth factors** and **stimulating factors** leading to different biochemical reactions determine the type of cells they will become. When these bio-chemicals are present within the environment of a colony of stem cells, they signal the cells to differentiate. Stem Cell Factor, Glycoprotein Growth Factor, and Colony Stimulating Factor are three different proteins that promote the differentiation process. These three proteins function in the presence of other biochemicals such as **transcription factors**, proteins that regulates and controls the transcription of DNA into RNA, determine the final cell type.

Knowledge and Comprehension:

Cell Differentiation:

Zygote:

Gene Expression:

Function:

Embryonic Stem Cells:

Growth factors and Stimulating factors:

Transcription factors:

1. Describe what cell differentiation is.

2. What happens to embryonic cells to transform them into different cell types?

Application, Analysis, Evaluation, and Synthesis

3. How is cell differentiation and gene expression related?

4. What experimental evidence demonstrates that embryonic cells from blastocysts can be grown and differentiate into different cell types. Explain how scientists achieved this in the laboratory.

5. Explain how growth factors and stimulating factors affect cells. Why are they important to differentiating cells?

The Evolution of Animals

Common Core Workbook

Monica Sevilla

Contents

What is Evolution?
What is an Adaptation?
What is Natural Selection?
Adaptive Radiation
Speciation
The Evolution of Fish
The Evolution of Amphibians
The Evolution of Reptiles
The Evolution of Mammals
The Evolution of Birds
The Evolution of Primates
The Transition from the Jungle to the Savanna

What is Evolution?

Evolution is a change in the characteristics or traits of organisms, such as plants, animals, and microorganisms the organism. Some beneficial changes, such as **adaptations**, help organisms adapt better an over a period of time. A change can occur in as little as a few generations. Every level of organization can be affected starting at the population level and moving down into the species, organism, and molecular levels.

The basis of evolutionary change occurs with the **DNA** (Deoxyribonucleic acid) molecules. DNA contains the instructions for the creation of living organisms on Earth. When **mutations** or changes in the DNA occur, they could be beneficial, harmful, or sometimes lethal to d live successfully within their environment. Harmful and lethal mutations impact the survival of an organism and may result in death. The death of an entire animal species is called **extinction**.

Diversity in the plant, animal, and microorganism species is due to the evolution of species throughout time. Scientists believe that all species on Earth were created 3.8 billion years ago from a single living organism, which changed and developed into many other animal species through the process of **speciation** throughout time. The evidence for this is the shared DNA sequences and physical traits. Traits are more similar in species that have a common ancestor. From this information, relationships between different species and their history of evolution can be created.

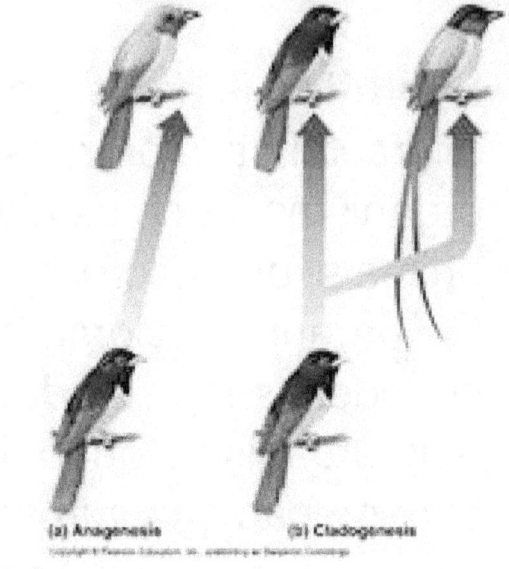

Source:

- Panno, Joseph (2005). *The Cell: Evolution of the First Organism*. Facts on File. ISBN 0-8160-4946-7. [page needed]

Knowledge and Comprehension
Words to Know:

Evolution:

DNA:

Mutations:

Adaptations:

Extinction:

Speciation:

1. What is evolution?

2. How is evolution related to DNA?

Application, Analysis, Evaluation and Synthesis

3. Explain how mutations can affect living organisms?

4. Why do some animal species go extinct?

5. Create an argument to support the claim "All the animal species on Earth evolved from a single animal species." Use evidence from the text to support your claim.

What is an Adaptation ?

An adaptive trait, or **adaptation**, is a trait that an organism has been kept, maintained, and has evolved over time.

These traits have **evolved** or changed through natural selection. **Natural selection** is a process by which members of a species who have inherited traits that give them a survival advantage that will eventually result in producing more offspring.

An adaptation is the result of an organism such as a plant or animal responding to the environment they live in. Organisms are continually responding or reacting to their **environment** or their surroundings. In order to survive successfully in the wild, organisms adjust to their environments by developing traits that will

give them a survival advantage. Nature is the selective agent which causes organisms to change or modify already existing traits. Adaptations can be either **physical** such as a structure on the body of an organism or **behavioral** which affects the way organism think, act, and behave. These adaptations can be inherited and passed down to offspring through the DNA.

The DNA within the cells of organisms can be affected by a wide variety of chemical reactions and physical reactions. Different components within the environment can affect the way these reactions progress within the cells. These affected reactions can cause **mutations** or changes in the DNA.

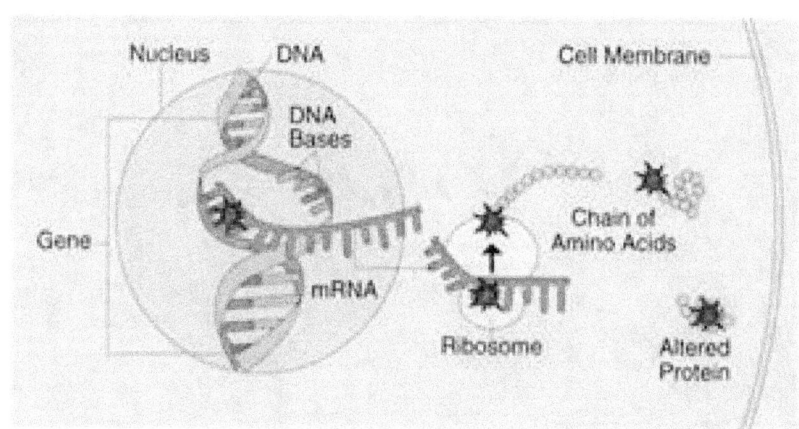

The mutation that occurs is a permanent change in the nucleotide sequence of the DNA within a gene. If the gene codes for a protein, this will affect the amino acid sequence of the protein it codes for and may change the structure and the function of the protein. These mutations are passed on to successive generations of offspring, increasing the number of individuals in a population who carry this mutation.

Knowledge and Comprehension:

Adaptation:

Evolve:

Natural Selection:

Environment:

Physical Adaptation:

Behavioral Adaptation:

1. What is an adaptation?

2. What is the process of natural selection.

Application, Analysis, Evaluation, and Synthesis

3. How does natural selection ensure the survivability of an organism?

4. What two types of adaptations can occur? How are they different?

5. Explain how the environment can cause adaptations to be passed down from generation to generation.

6. Why are adaptations important to organisms? Support your response by using evidence from the text.

What is Natural Selection?

Charles Darwin, in 1859, published his famous book " On the Origin of Species by Natural Selection" proposed the theory of the evolution of plants and animals by a process called natural selection. **Natural selection** is the process by which plants and animals that have a better ability to adapt to their environment survive and reproduce with more success than other plants and animals.

There are 4 different factors that affect natural selection: overproduction, inherited variation, struggle to survive, and successful reproduction.

Overproduction: some animals that have been reproduced will survive and some of them will not. Animals will tend to have many offspring to ensure that some will survive.

Inherited Variation: every individual has its own genetic makeup that is a combination of the genetic material passed down to them during reproduction. Each individual has is similar, but not identical to its parents.

Survival: Some animals may not survive to adulthood because of predators, diseases, or starvation. Only a few of these animals will survive.

Reproduction: The animals that are best adapted to their environment are likely to survive and reproduce.

Charles Darwin did not know how genetic material was passed on from one generation to the next. He did know that there were many **variations** or different types of traits within animals. **Traits** are the characteristics of an individual. He concluded that the variation in traits was somehow determined by this genetic material which we now know as DNA. He also concluded that this genetic material was passed down from parents to their offspring during reproduction and that some of the inherited traits in the offspring give them an advantage for survival.

Knowledge and Comprehension
Words to Know:

Natural Selection:

Overproduction:

Inherited variation:

Survival:

Reproduction:

Variation:

Trait:

1. What is natural selection?

2. What 4 factors affects natural selection?

Application, Analysis, Evaluation and Synthesis

3. How would individual traits within a small group of animals be affected if all the animals were destroyed in a forest fire?

4. If many individuals in a population of rabbits reproduced successfully, a passed on a trait that give them a survival advantage, how would this affect the rabbits?

5. Explain how traits found in parents can give their offspring a survival advantage.

Adaptive Radiation

Adaptive radiation is a process of evolution in which organisms **diversify** or develop into many new forms. This rate a which this occurs is usually rapid. Adaptive radiation is commonly caused by new environments that have recently become available, new resources within an environment become available, or organisms have new challenges that they must overcome.

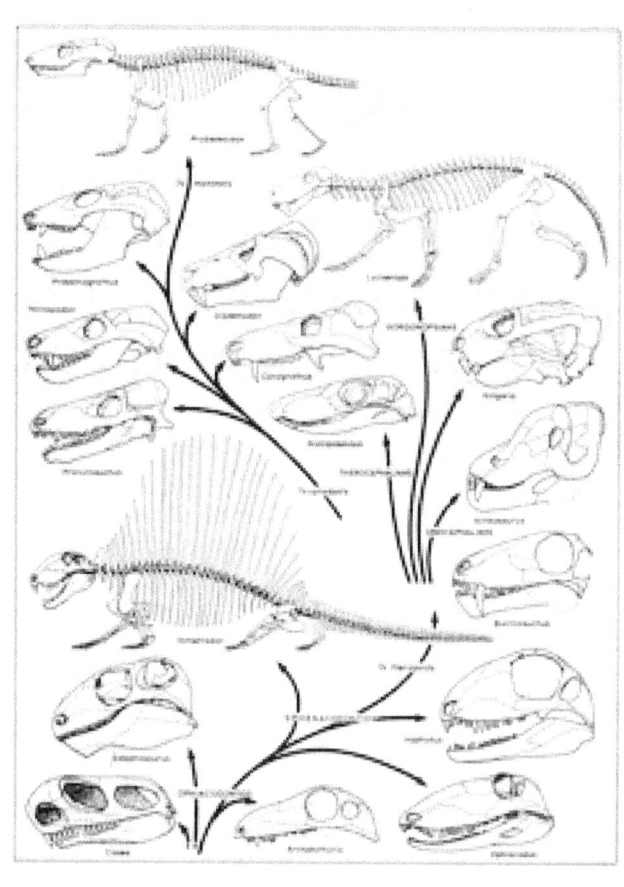

Adaptive radiation causes **speciation** or the creation of new species to occur over time. This is the result of one species diverging or separating into a new, unique species. A **new species** forms when it can no longer mate with the species it separated from. The sequences of their DNA

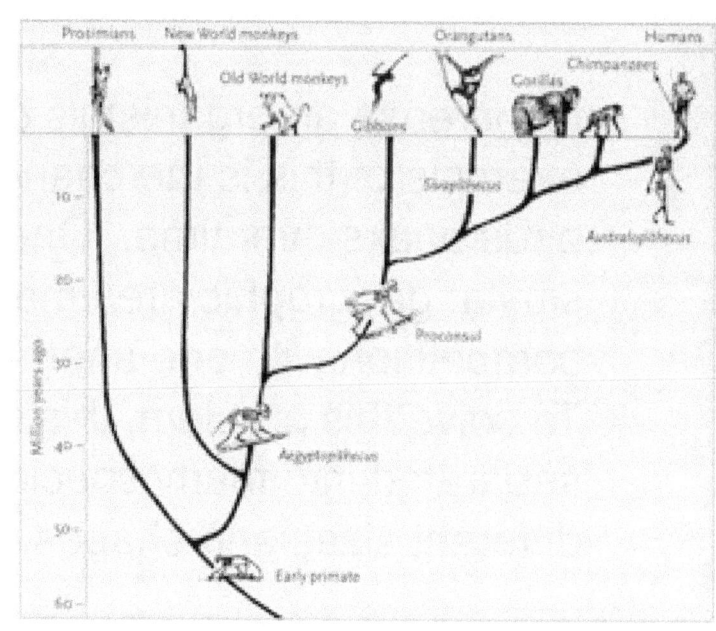

(Deoxyribonucleic Acid) become different enough that they are unable to produce viable and fertile offspring. This can occur if two populations from the same species are isolated from one another either geographically or reproductively. The genetic material, in this case, is not exchanged between the two populations. This concept is known as **gene flow**. Gene flow, in this case is restricted between the two populations. If enough time passes, each population will become genetically different from one another.

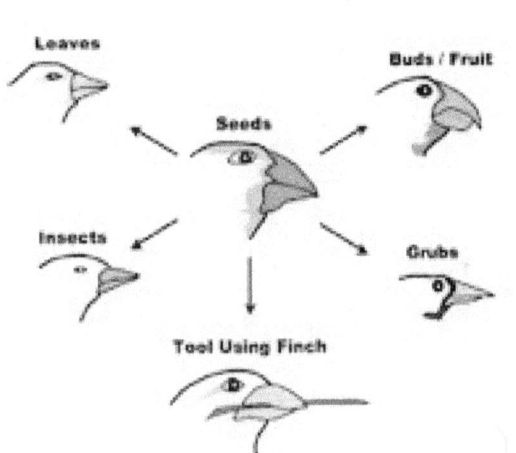

Adaptive radiation also causes phenotypic changes or changes in the an organism's **traits** to occur. Some of these adaptations may affect specific structures of the body of an organism. An **adaptation** is a change that is made to the structure or function of the body to increase an organism's chances for survival. An example of this is the change in the size and shape of birds beaks over time. One species of birds inhabiting a similar geographic area may have been in extreme competition with one another for the same food resources. To solve this problem, this species may have diverged or separated into many species of birds with beaks of different sizes and shapes. This could have occurred out of a need to exploit new resources within a geographic area.

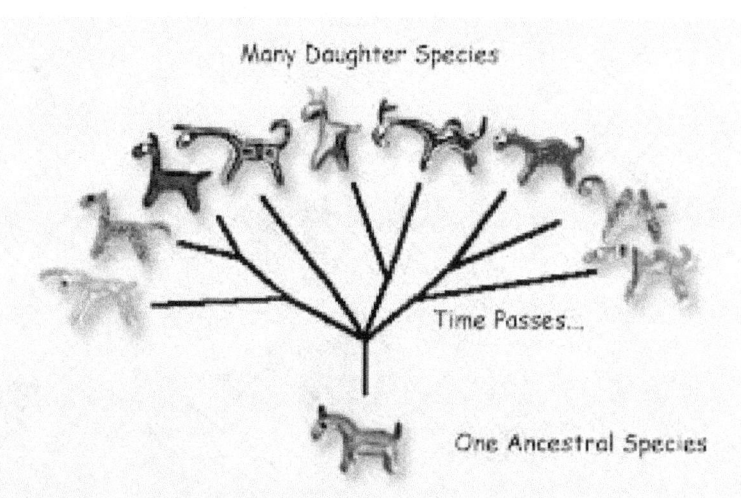

This scenario is what Charles Darwin observed on the Galapagos Islands in South America. He found that the finches, now known as **Darwin's finches**, that inhabited the island had different heads and different beaks. He concluded, based on his observations, that many different finches, who specialized in eating different food sources, evolved from one ancestral seed-eating ground finch. Some of these finches ate seeds, some ate buds and fruits, some ate cactus seeds, and others ate insects. He theorized that the members of ancestral species of finches were all competing for the same food source (seeds) at one time, and they all found different food sources on different parts of the islands. This took them out of direct **competition** with each other, and over time, the finches developed into different daughter species. Each species evolved to have heads and beaks with different shapes and sizes. Each beak adapted and specialized to the type of food they were eating at the time of Darwin's observations.

Adaptive radiation in Galapagos finches

- medium tree finch (*Camarhynchus pauper*)
- large tree finch (*Camarhynchus psittacula*)
- small tree finch (*Camarhynchus parvulus*)
- mangrove finch (*Camarhynchus heliobates*)
- vegetarian finch (*Camarhynchus crassirostris*)
- woodpecker finch (*Camarhynchus pallidus*)
- large cactus finch (*Geospiza conirostris*)
- warbler finch (*Certhidea olivacea*)
- cactus finch (*Geospiza scandens*)
- Cocos Island finch (*Pinaroloxias inornata*)
- sharp-beaked ground finch (*Geospiza difficilis*)
- small ground finch (*Geospiza fuliginosa*)
- large ground finch (*Geospiza magnirostris*)
- medium ground finch (*Geospiza fortis*)

Inner ring categories: buds and fruits; mainly insects; cactus seeds and parts; mainly seeds. Center: ancestral seed-eating ground finch.

© 2005 Encyclopædia Britannica, Inc.

Knowledge and Comprehension:

Adaptive Radiation:

Diversified:

Adaptation:

Speciation:

New Species:

Gene Flow:

Traits:

Darwin's Finches:

Competition:

1. Describe what adaptive radiation is.

2. What does adaptive radiation cause over time?

3. Explain how a new species can develop over time.

Application, Analysis, Evaluation, Synthesis

4. Explain how competition can cause a new species to occur.

5. Explain how Darwin came to the conclusion that the finches he observed from the Galapagos Islands all had the same ancestor species.

Speciation

Over the entire history of the Earth, many different forms of life have existed. It is estimated by scientists that 96% of the species that ever existed on the Earth, have gone **extinct** or are no longer living. A **species** is defined as the largest group of organisms that are able to interbreed and have fertile offspring. Every major extinction in Earth's geologic history has given rise to new species.

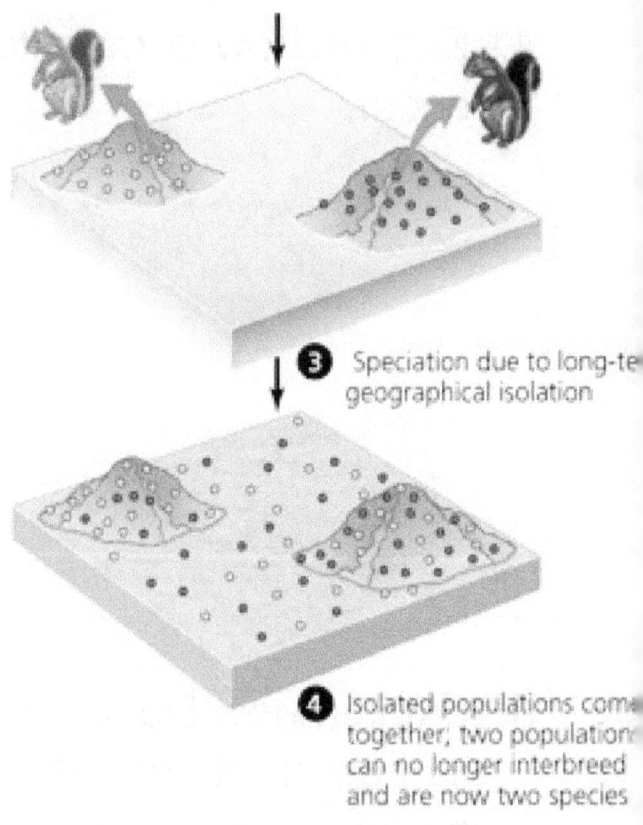

3. Speciation due to long-term geographical isolation

4. Isolated populations come together; two populations can no longer interbreed and are now two species

Barriers to reproduction may arise that reduce or prevent the gene flow between populations of the same species. **Gene flow** is the transfer of alleles from one population to another through time. Barriers can cause new alleles within a species to develop. **Alleles** are simply the variation of traits that are expressed by a gene.

Reproductive isolation occurs through a specific set of processes and behaviors that reduce or prevent the members of the same species from reproducing and having offspring. Sometimes this happens through the geographical separation of one **population** from another. Climate change or the destruction of a habitat can cause the resources within a habitat to change and become scarce. This may be an important cause for a species to break up into different groups or populations to seek more suitable locations to survive in. As a result, **speciation**, the process which allows 2 or more new species to appear, may occur.

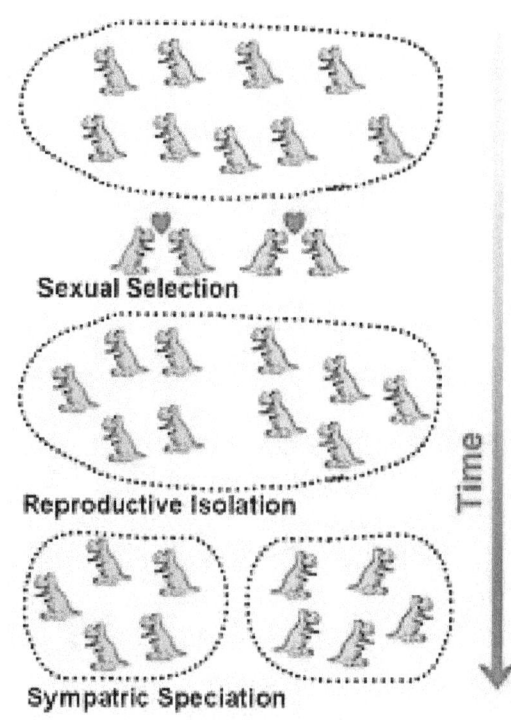

New species of animals or plants will not be able to reproduce with each other and leave fertile offspring. This is because their DNA has become significantly different from each other. The gene flow between the original species and the old species is disrupted. A new species can form over many generations.

Sources:

http://en.wikipedia.org/wiki/Species
http://evolution.berkeley.edu/evolibrary/article/evo_42

Knowledge and Comprehension:

Extinct :

Species:

Gene Flow:

Alleles:

Reproductive Isolation:

Population:

Speciation:

1. Describe what a species is.

2. What is speciation?

3. Explain how speciation can occur.

Application, Analysis, Evaluation, Synthesis

4. Explain how reproductive isolation can cause speciation to occur.

5. Explain why new species cannot mate with the original species.

The Evolution of Fish

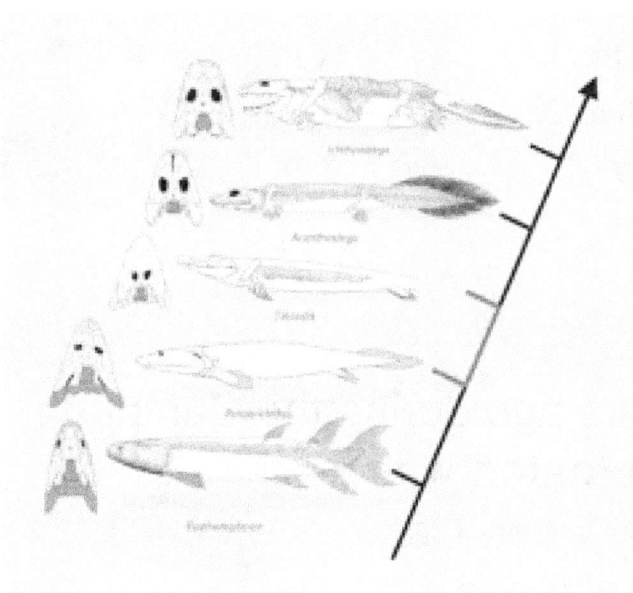

Evolution of fish as a life form is another important development in the history of life on Earth. Fish were the first time vertebrate species were created. **Vertebrates** are animals that have a backbone or a spine. Before this, **invertebrate** species, animals without a backbone or spine, such as the Crustaceans, hard-shelled animals such as crabs and lobsters, were inhabiting our oceans along with many plant, single-celled organisms, and other multicellular organisms.

Hard-shelled animals such as corals and star fish were able to construct their shells, the hard, protective, outer covering of some organisms, through the use inorganic compounds such as calcium carbonate. The compound calcium carbonate is a product of biomineralization. **Biomineralization** is the process of an animal uses to make their own complex molecules and minerals from chemical elements present on Earth. Early animals were able to take molecules such as calcium, carbon dioxide, present in the ocean water, to synthesize or make their shells. Scientists believe that they may have evolved from a sea squirt. These soft bodied animals adapted to their environment through the formation of a backbone, as well as bones. Bones are also formed through the process of biomineralization.

Fish originated 530 million years ago during the Cambrian explosion. The **Cambrian explosion** was a period of time in Earth's history when life **diversified**. This means that all types of different life forms appeared. This was the first time, however, that vertebrates or animals with a backbone appear on the Earth. The first fish that formed was a jawless, armored fish known as a **Ostracoderm**.

The first fish with jaws were known as a **Placoderm**. The jaw was an adaptation in fish that gave them an evolutionary advantage over jawless fish. Jaws improved the respiration of fish, as well as, their force of biting, improving their selection of food sources. Fish with jaws eventually evolved into amphibians.

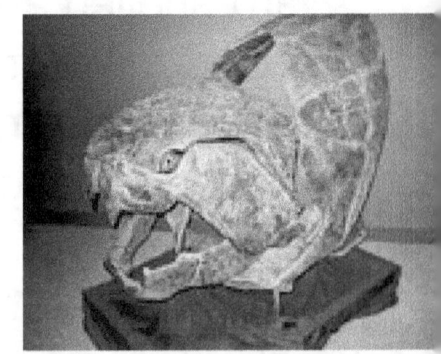

Knowledge and Comprehension:

Vertebrates:

Invertebrates:

Biomineralization:

Cambrian Explosion:

Diversified:

Osracoderm:

Placoderm:

1. What is a vertebrate?

2. What is the major difference between vertebrates and invertebrates?

3. Which animal do scientists believe that fish evolved from?

Application, Analysis, Evaluation, Synthesis

4. Explain what biomineralization is and how it is related to animals that have bones.

5. During what period of time were fish created? Why is the evolution of fish important?

6. How did fish adapt to their ocean environment? What adaptation was made?

7. What animal did fish evolve into? Do you agree with this? Explain your reasoning.

The Evolution of Amphibians

Evolution of amphibians marked the first time animals with lungs capable of breathing air evolved on land. Amphibians are members of animals known as **vertebrates** or animals that have a backbone or a spine. They evolved from fish that developed jaws. The first fish with jaws were known as a **Placoderm**. The jaw was an adaptation that gave fish an evolutionary advantage for survival. Jaws improved the respiration of fish, and their ability to bite, improving their ability to eat and their selection of food sources.

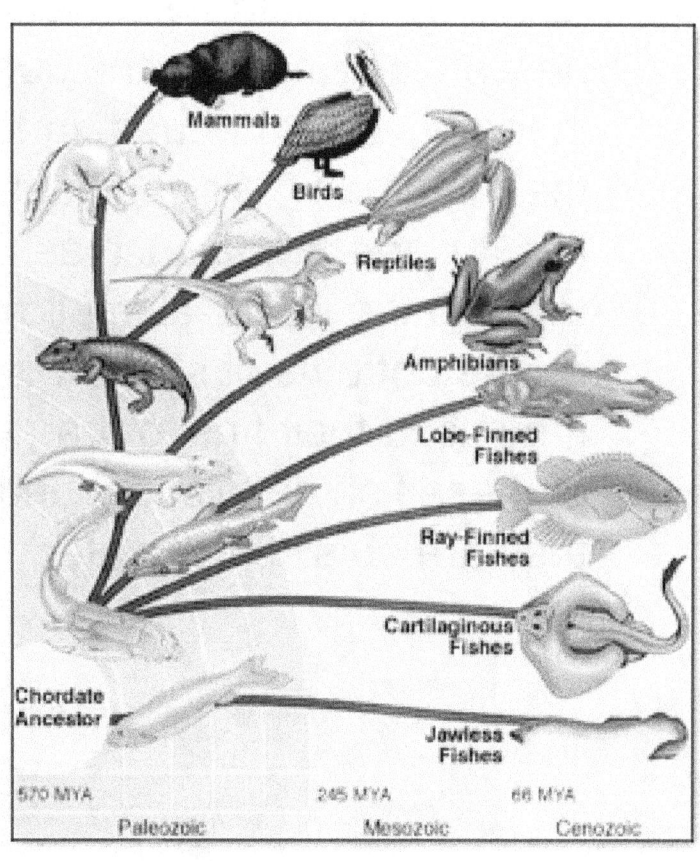

Scientists believed amphibians evolved specifically from a species of fish known as the **lobe-finned fish.** This fish once lived in the ocean, during the Devonian Period,

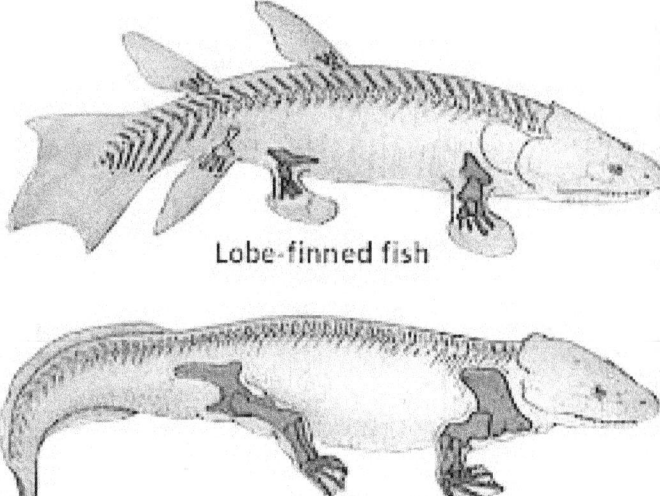

Lobe-finned fish

Early amphibian

350 million years ago. They made a few important **adaptations** or changes to their bodies, that allowed them to live on land and allow early amphibians to dominate a new environment on the surface of the Earth, rather than in the oceans. Lobbed finned fish had hard, bony fins on the bottom of their bodied that evolved into legs. Their fins, **appendages** such as limbs that allowed fish to swim through the water, also allowed them to drag themselves on the bottoms of shallow seas and shores. Eventually, they used their fins to drag themselves onto land. Some of these fish developed lungs to be able to breath air while they are outside the water. They still, however, lay their

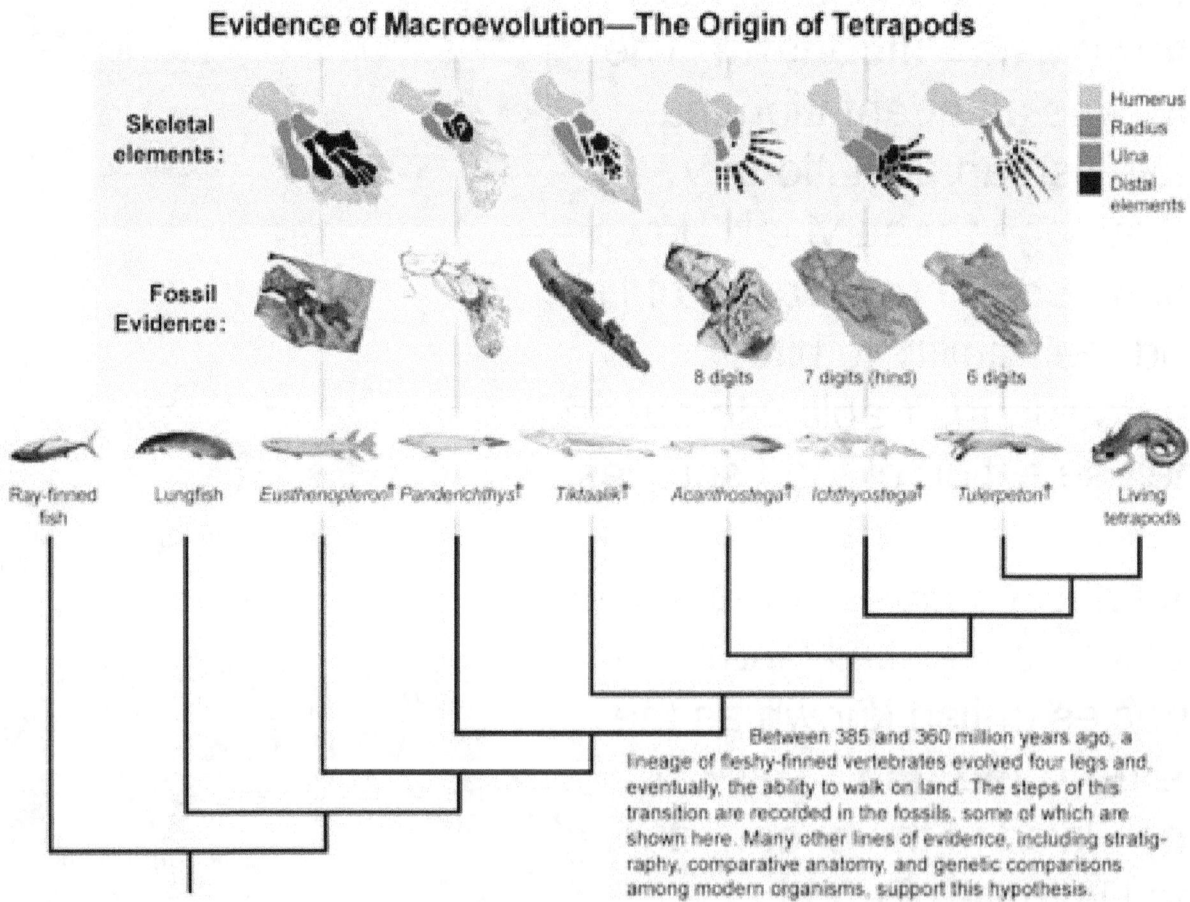

eggs in the water and their tadpoles remain in the water until they develop fully functioning lungs and legs. This was the first time in the history of the Earth that **tetrapods**, or animals with four legs, appeared. The lobbed fish is the ancestor of all modern land animals today.

Knowledge and Comprehension:

Vertebrates:

Placoderm:

lobe finned fish:

Adaptations:

Appendages:

Tetrapods:

1. What is a tetrapod?

2. What do fish use fins for?

3. Which fish species evolved into the first amphibians?

Application, Analysis, Evaluation, Synthesis

4. Explain how fins were adapted into legs. Why is this adaptation important for land animals?

5. What important adaptations did amphibians need to make in order to be able to live on land?

6. Explain why you think fish needed to move out of the water and onto land.

The Evolution of Reptiles

Evolution of reptiles occurred during the Carboniferous Period in Earth's history about 310-320 million years ago. They evolved or changed from amphibians that lived in and around swamps. These amphibians became land-based as a result of heavy competition for food with other animals. To exploit and take advantage of distant areas that offered undisturbed resources, such as plants and insects, they spent increasingly more time of land than in the water.

Reptiles evolved differently than their amphibian ancestors in a few important ways. Animals **evolve** or change in response to their environment. **Adaptations**, structural changes in the body, allow animals to improve their chances of survival in their environment. To make a full transition from the water to the land, they adapted by forming hard-shelled eggs that they could lay on land. These eggs contained the developing fetus surrounded by an amniotic membrane that keeps

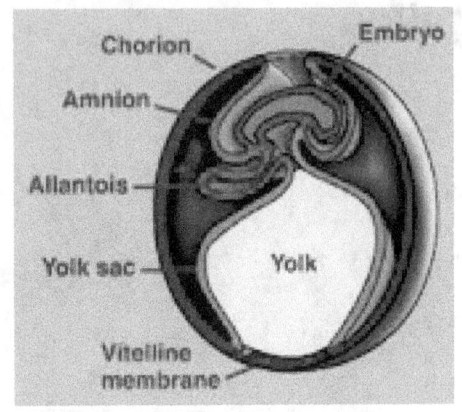

fluids and nutrients inside the shell. For this reason, reptiles are known as **amniotes**. Animals that develop amniotic membranes to protect and nourish their fetuses. Reptiles are the Earth's first amniotes.

The skin of reptiles changed from a smooth texture to one with scales. Scales helped early reptiles to adapt to an environment surrounded by air rather than water, and helped to protect them from the elements such as the sun, the wind and the rain. Some of them also acquired to the ability to molt or shed their dry and damaged skin.

Reptiles also have bigger brains than amphibians. Their cerebrum and cerebellums are larger in size. This increase in brain size allowed reptiles to develop improved hunting strategies, an increase in the development of their senses, and an increase in the motor control of their muscles.

Reptiles are cold blooded. They cannot control their internal temperature. They rely on natural sources of heat that can be found in their surroundings. They will

bask in the sun on top of a rock if they need to warm up or burrow underground to cool down.

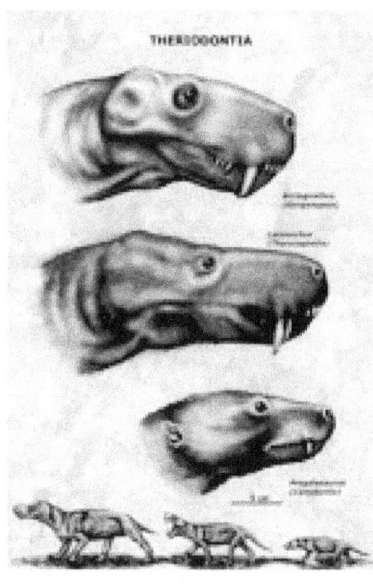

Mammals, animals that are warm-blooded, have hair, mammary glands, and placental births, evolved from reptiles. The reptiles they evolved from were known as **Therapsids** or mammal-like reptiles. These reptiles had many of the characteristics that mammals have today. They evolved during the Permian.

The reptilian ancestor of the therapsids was a dinosaur known as a **Pelycosaur**. Reptiles also evolved into birds which are also warm-blooded.

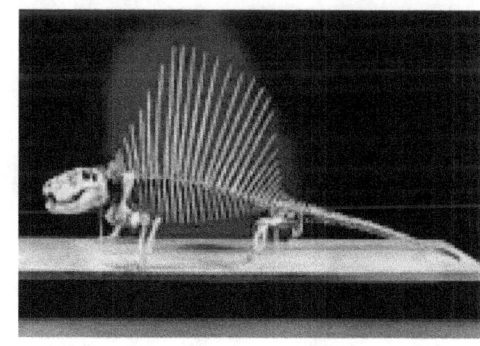

Knowledge and Comprehension:

Evolve:

Adaptation:

Amniotes:

Mammals:

Therapsids:

Pelycosaur:

1. What are the characteristics of a reptile?

2. Which reptile species evolved into mammals? Who was its ancestor?

3. Reptiles evolved into what two groups of animals? What is similar about them?

Application, Analysis, Evaluation, Synthesis

4. What are some ways that reptiles adapted to living on land?

5. Explain why the reptiles adapted to living on land.

6. In your opinion, what characteristics did the ancestor of the mammals, the Therapsid, have? Explain why you think so.

The Evolution of Mammals

Evolution of mammals occurred during the **Carboniferous Period** in Earth's history with the appearance of a group of reptiles known as the **Synapsids**. Synapsids evolved into two evolutionary lines around 260 million years ago during the Permian Period, around 298 million years ago. One of these groups are **Therapsids** that became known as mammal-like retiles that evolved into **amniotes** or animals that could lay hard-shells on land. The shells developed amniotic membranes to protect and nourish their fetuses. The

amniotic membrane keeps fluids and nutrients inside the shell and protects them from them from the surrounding environment. By the middle of the Triassic period, there were many Therapsid species that looked like mammals and had the characteristics of mammals. **Mammals,** animals that are warm-blooded, have hair, mammary glands, and placental births, evolved from reptiles.

One specific Therapsid species, *Procynosuchus delaharpeae,* has been identified by scientists to have evolved into today's mammals. This species lived in South Africa during the Late Permian. It is one of the earliest cynodonts that have been identified. A **cynodont** is a group of animals and their descendants that evolved into today's mammal species, including humans. Cynodonts, that are non-mammalian, did not remain in South Africa, but spread throughout the southern part of Gondwana. **Gondwana,** one of the two landmasses present on Earth at this time, was composed of the continents of Africa, India, South America, and Antarctica. Evidence of this are fossils of these species that have been found in these continents. Fossils also have been found in eastern North America and Western Europe (northwestern France and Belgium). These fossils suggest that non-mammalian cynodonts migrated throughout Pangaea, while the cynodonts that became mammals, and eventually humans, remained in South Africa until after the break up of **Pangaea**, a single landmass and a supercontinent, into the two landmasses of Gondwana and Laurasia.

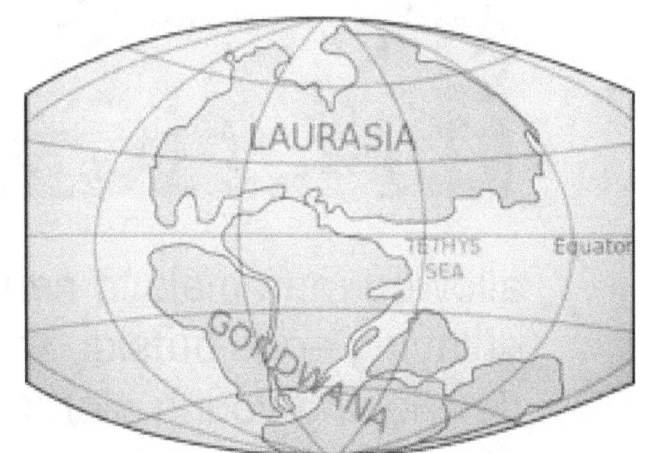

TRIASSIC
200 million years ago

Mammals with placentas did not evolve until during the Cretaceous period about 100 to 85 million years ago. Placentas allowed these animals to carry and develop their fetuses within their bodies. This adaptation allowed mammals to carry and protect their young instead of laying eggs outside the body and leaving their developing young alone and unattended while they forage for food. Primates evolved at the end of the Cretaceous period. Mammals were small in size when they first evolved during this time but grew bigger during the early Paleocene. Around the same time, the area of the brain in control of smell in mammals began to enlarge. This **adaptation**, or change in the structure brain structure and function in response to the environment, helped them improve their hunting strategies in the night during the nighttime when large dinosaurs were asleep. Since they could not rely on sight for catching prey animals in the dark, they developed a better sense of smell to compensate for this.

Knowledge and Comprehension:

Carboniferous Period:

Synapsids:

Therapsids:

Amniotes:

Mammals:

Cyanodont:

Gondwana:

Pangaea:

1. Describe what a Synapsid is.

2. How are Synapsids and Therapsids different?

3. What is an amniote?

4. What adaptation did the first mammals make to improve hunting strategies during the night?

Application, Analysis, Evaluation, Synthesis

5. What happened to the non-mammalian cynodonts? Explain why you think mammalian cynodonts evolved in South Africa and not anywhere else on the Earth.

6. Fossils of non-mammalian cynodonts have been found in South Africa to South America, North America, and Europe. If they were first created and formed in South Africa, explain why they are found on different continents?

7. Explain why the animal species *Procynosuchus delaharpeae,* is important. Support your answer with evidence from the text.

The Evolution of Birds

Reptiles first evolved during the Carboniferous Period in Earth's history about 310-320 million years ago. They evolved from amphibians whose ancestor was the lobe finned fish. Birds evolved from a group of specialized dinosaurs known as **Therapoda** or Therapods. They evolved 230 million years ago during the **Triassic period**.

Therapoda were dinosaurs that were **bipedal**, or walked on 2 legs, had small, bumpy scales, were covered with feathers or feather-like structures, were carnivores, and looked like lizards. These dinosaurs shared characteristics of birds such as a wishbone, incubation of eggs, and bones filled with air. One of the species of Therapoda, known as **Coelurosaur**, evolved and developed feathers. All modern birds evolved from Coelurosaurs about 66 million years ago during the Cretaceous period.

Birds like many organisms before them, made many adaptations in response to their changing environments. These **adaptations**, or changes in their structures in response to their environment, include: the formation of a beak, wings, and a reduction in size. Their reptile ancestors already had hollow bones, feathers, claw-like fingers, a snout with teeth, and laid eggs. It is clear, especially in the case of birds, that many of their hallmark characteristics were adaptations that were developed millions of years earlier in their reptile ancestors.

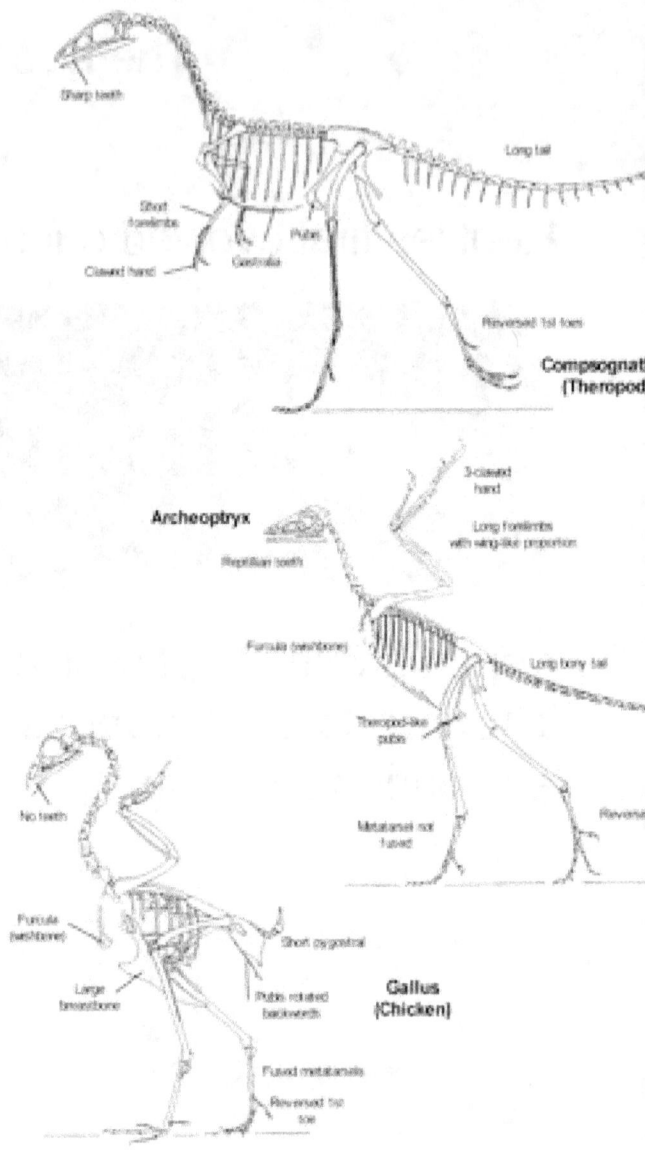

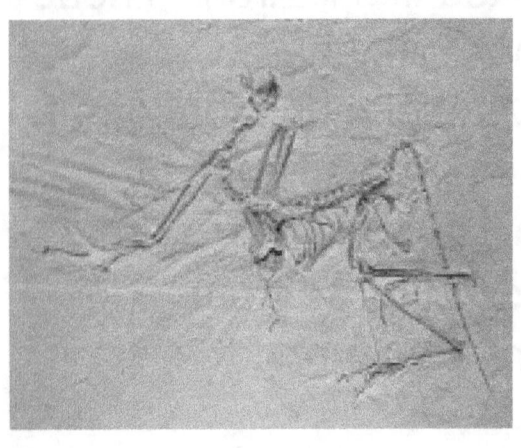

Archeopteryx, a bird that lived during the late Jurassic period about 150 million years ago, is known as the oldest member of the group Avialae or Aves. This species has been accepted by scientists as being the first true

bird that evolved on Earth. It had feathers, it was warm blooded, it had wings, it had two legs, and laid eggs.

Knowledge and Comprehension:

Therapoda:

Triassic Period:

Coelurosaur:

Adaptations:

Archeopteryx:

Bird:

1. Describe the characteristics of a bird.

2. Why are the members of the group Therapoda important?

3. Which characteristics are similar in Therapoda and modern birds?

Application, Analysis, Evaluation, Synthesis

4. When birds were evolving for the first time, which of their characteristics came from reptiles?

5. As birds evolved, did they already have all the hallmark characteristics that birds have? Why or why not? Support your answer with evidence from the text.

Primate Evolution

The mass extinction of the dinosaurs opened the door for the evolution of a new type of mammal, the primate. The appearance of primates occurred roughly 85,000 years ago, during the Cretaceous period. **Primates** evolved from mammals that lived in trees and evolved flexible hands and digits to be able to swing and move through the tropical and subtropical rainforests of Africa. They split, millions of years ago, into two important types of primates: prosimians and simians. Each type evolved their own characteristics as they adapted to their environments.

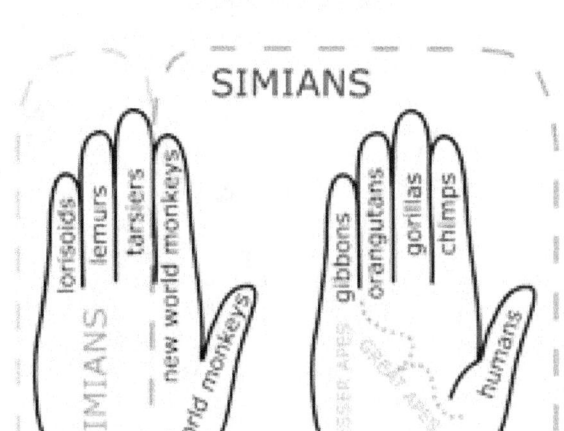

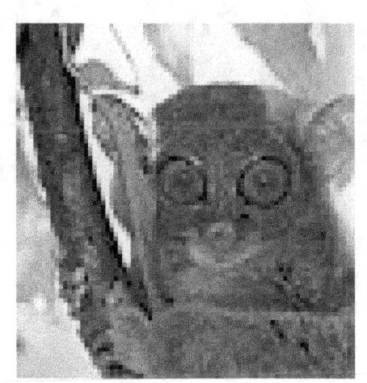

Prosimians such as lemurs and tasiers, are the oldest living simians on our planet. They are small, nocturnal creatures. The are mostly tree dwellers and hunt for small animals and insects for food. They are considered to have characteristics that are primitive when compared to the simians. They lack color vision. Their eyes maximize the number of rods in their retinas. This allows for better vision in conditions with low lighting. They have small brains when compared to the simians.

Simians, also known as anthropoids, are a group of primates that include new world monkeys, old world monkeys, lesser apes such as gibbons, and the great apes such as orangutans, gorillas, chimpanzees, the hominids, and humans. The simians are classified into three separate groups: old world monkeys, new world monkeys, and apes.
humans and the hominids are included as apes. The new world monkeys diverged from old world monkeys and the apes about 25 million years ago.

The characteristics of simians have evolved as adaptations that helped primates transition from one environment to another. **Adaptations** are features that help organisms to survive in their environment. The ancestors of the simians moved from the trees into the grasslands and the savannas in Africa. This may have been due to the availability of new food sources in

these environments. As these primates transitioned to a new way of life, they became **bipedal**, walking on two legs, hunted during the day, and learned how to use tools to survive. Their brain sizes, compared to their prosimian ancestors, became bigger and more sophisticated throughout time. Some of the simians have developed color vision, as well as **stereoscopic vision** and depth perception, which are the result of having both eyes in the front of the head.

Chimpanzees, gorillas, as well as humans, share similar characteristics characteristics. They all demonstrate the similar social behaviors and emotions. Gorillas and chimpanzees have demonstrated that they could communicate with humans through sign language. The IQ of chimpanzees can reach up to 80, the average IQ of a human and it has also been proven that human DNA is 98% similar to chimpanzee DNA.

Knowledge and Comprehension
Words to Know:

Primates:

Prosimians:

Simians:

Adaptation:

Bipedal:

Stereoscopic Vision:

1. How long ago did primates evolve?

2. Describe the characteristics of the prosimians.

3. Describe the characteristics of the simians.

Application, Analysis, Evaluation and Synthesis

3. Explain how prosimians and simians are different when compared to each other.

4. Explain how and why simians evolved from their prosimian ancestors.

5. Do you agree or disagree with the claim "Humans, gorillas, and chimpanzees demonstrate similar characteristics." Use evidence from the text to support your claim.

The Transition from the Jungle to the Savanna

The first primates appeared roughly 85,000 years ago during the Cretaceous period. **Primates** evolved from mammals that lived among trees of the tropical jungles and subtropical rainforests of Northern Africa. They split, millions of years ago, into two important types of primates: prosimians and simians. The **prosimians** were primarily tree dwellers. Their food sources, small animals and insects, were located within the trees. A drastic change in the climate occurred causing a world-wide cooling trend.

This cooling trend caused the average global temperature to drop by 10 to 20 degrees Fahrenheit, allowing the onset of an ice age. As the climate changed, the trees in the jungle biome slowly disappeared. These open areas turned into savannas and grasslands.

The ancestors of the great apes and the hominids, the early **simians**, adjusted to their new environment in the

savannas by developing adaptations that help organisms to survive in a new environment. These adaptations, becoming bipedal walking on two feet, and use of tools caused the brains of these primates to reorganize and to grow in size. These early changes in locomotion freed the hands and allowed them to concentrate on making and using tools.

Creating and using tools mentally challenged early primates. This was a complex mental process that required the early primates to understand the purpose for making a tool, envision the finished product they wanted to create, and figure out different techniques of making the tool. Through the use of the tool, they could evaluate it and modify it to improve its performance.

Knowledge and Comprehension
Words to Know:

Primates:

Prosimians:

Simians:

Adaptation:

Bipedal:

1. When did the ancestors of the hominids transition from the jungle to the savanna.

2. Explain why the ancestors of the hominids were forced to move into the savannas.

3. What major adaptation did the ancestors of the hominids make as they adjusted to living in the savannas?

Application, Analysis, Evaluation and Synthesis

4. How did bipedalism, walking on 2 feet, help the hominids develop the use of tools.

5. Explain how new adaptations the hominids made initiated the reorganization of the brain and the growth of the brain.

6. Create a short storybook about how the ancestors of the hominids transitioned from the jungles into the savannas. Include pictures along with your writing.

Common Core Biology:

Human Evolution

Monica Sevilla

Contents

What is Evolution?

Primate Evolution

What are Adaptations?

The Transition from the Jungle to the Savanna

The Transitional Species: Homo Habilis and Homo Rudolfensis

Homo Erectus

Homo Heidelbergensis

Homo Neanderthalensis

Differences in Adaptations among Neanderthals and Homo Sapiens

The Denisovans

Homo Floresiensis

Coexistence: Homo Sapiens and Other Hominids

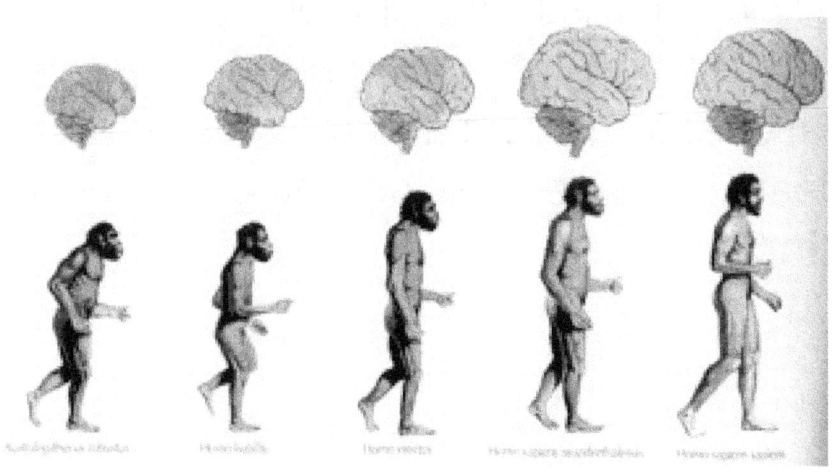

What is Evolution?

Evolution is a change in the characteristics or traits of organisms, such as plants, animals, and microorganisms the organism. Some beneficial changes, such as **adaptations**,  help organisms adapt better an over a period of time. A change can occur in as little as a few generations. Every level of organization can be affected starting at the population level and moving down into the species, organism, and molecular levels.

The basis of evolutionary change occurs with the **DNA** (Deoxyribonucleic acid) molecules. DNA contains the instructions for the creation of living organisms on Earth. When **mutations** or changes in the DNA occur, they could be beneficial, harmful, or sometimes lethal to d live successfully within their environment. Harmful and lethal mutations impact the survival of an organism and may result in death. The death of an entire animal species is called **extinction**.

Diversity in the plant, animal, and microorganism species is due to the evolution of species throughout time. Scientists believe that all species on Earth were created 3.8 billion years ago from a single living organism, which changed and developed into many other animal species through the process of **speciation** throughout time. The evidence for this is the shared DNA sequences and physical traits. Traits are more similar in species that have a common ancestor. From this information, relationships between different species and their history of evolution can be created.

(a) Anagenesis (b) Cladogenesis

Source:

- Panno, Joseph (2005). *The Cell: Evolution of the First Organism*. Facts on File. ISBN 0-8160-4946-7. [page needed]

Knowledge and Comprehension
Words to Know:

Evolution:

DNA:

Mutations:

Adaptations:

Extinction:

Speciation:

1. What is evolution?

2. How is evolution related to DNA?

Application, Analysis, Evaluation and Synthesis

3. Explain how mutations can affect living organisms?

4. Why do some animal species go extinct?

5. Create an argument to support the claim "All the animal species on Earth evolved from a single animal species." Use evidence from the text to support your claim.

Primate Evolution

The mass extinction of the dinosaurs opened the door for the evolution of a new type of mammal, the primate. The appearance of primates occurred roughly 85,000 years ago, during the Cretaceous period. **Primates** evolved from mammals that lived in trees and evolved flexible hands and digits to be able to swing and move through the tropical and subtropical rainforests of Africa. They split, millions of years ago, into two important types of primates: prosimians and simians. Each type evolved their own characteristics as they adapted to their environments.

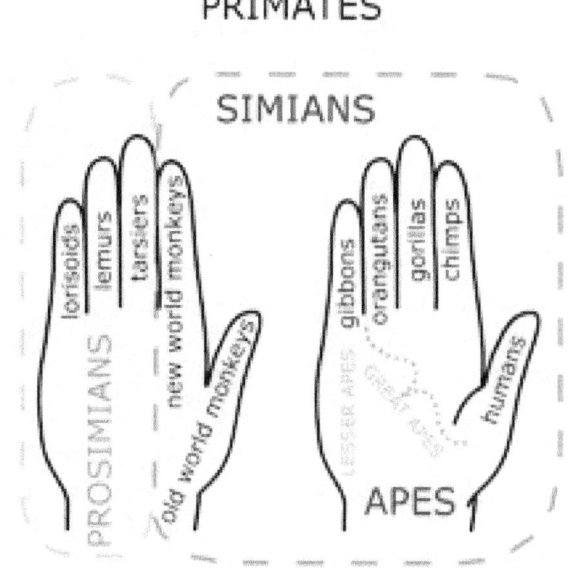

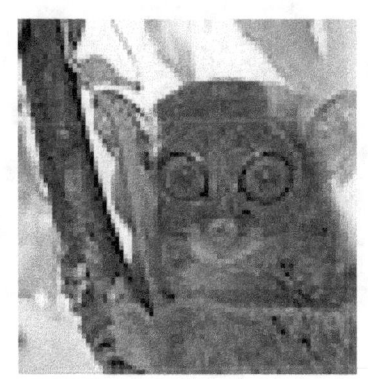

Prosimians such as lemurs and tasiers, are the oldest living simians on our planet. They are small, nocturnal creatures. The are mostly tree dwellers and hunt for small animals and insects for food. They are considered to have characteristics that are primitive when compared to the simians. They lack color vision. Their eyes maximize the number of rods in their retinas. This allows for better vision in conditions with low lighting. They have small brains when compared to the simians.

Simians, also known as anthropoids, are a group of primates that include new world monkeys, old world monkeys, lesser apes such as gibbons, and the great apes such as orangutans, gorillas, chimpanzees, the hominids, and humans. The simians are classified into three separate groups: old world monkeys, new world monkeys, and apes.
humans and the hominids are included as apes. The new world monkeys diverged from old world monkeys and the apes about 25 million years ago.

The characteristics of simians have evolved as adaptations that helped primates transition from one environment to another.
Adaptations are features that help organisms to survive in their environment. The ancestors of the simians moved from the trees into the grasslands and the savannas in Africa. This may have been due to the availability of new food sources in

these environments. As these primates transitioned to a new way of life, they became **bipedal**, walking on two legs, hunted during the day, and learned how to use tools to survive. Their brain sizes, compared to their prosimian ancestors, became bigger and more sophisticated throughout time. Some of the simians have developed color vision, as well as **stereoscopic vision** and depth perception, which are the result of having both eyes in the front of the head.

Chimpanzees, gorillas, as well as humans, share similar characteristics characteristics. They all demonstrate the similar social behaviors and emotions. Gorillas and chimpanzees have demonstrated that they could communicate with humans through sign language. The IQ of chimpanzees can reach up to 80, the average IQ of a human and it has also been proven that human DNA is 98% similar to chimpanzee DNA.

Knowledge and Comprehension
Words to Know:

Primates:

Prosimians:

Simians:

Adaptation:

Bipedal:

Stereoscopic Vision:

1. How long ago did primates evolve?

2. Describe the characteristics of the prosimians.

3. Describe the characteristics of the simians.

Application, Analysis, Evaluation and Synthesis

3. Explain how prosimians and simians are different when compared to each other.

4. Explain how and why simians evolved from their prosimian ancestors.

5. Do you agree or disagree with the claim "Humans, gorillas, and chimpanzees demonstrate similar characteristics." Use evidence from the text to support your claim.

What is an Adaptation ?

An adaptive trait, or **adaptation**, is a trait that an organism has been kept, maintained, and has evolved over time.

These traits have **evolved** or changed through natural selection. **Natural selection** is a process by which members of a species who have inherited traits that give them a survival advantage that will eventually result in producing more offspring.

An adaptation is the result of an organism such as a plant or animal responding to the environment they live in. Organisms are continually responding or reacting to their **environment** or their surroundings. In order to survive successfully in the wild, organisms adjust to their environments by developing traits that will

give them a survival advantage. Nature is the selective agent which causes organisms to change or modify already existing traits. Adaptations can be either **physical** such as a structure on the body of an organism or **behavioral** which affects the way organism think, act, and behave. These adaptations can be inherited and passed down to offspring through the DNA.

The DNA within the cells of organisms can be affected by a wide variety of chemical reactions and physical reactions. Different components within the environment can affect the way these reactions progress within the cells. These affected reactions can cause **mutations** or changes in the DNA. The mutation that occurs is a permanent change in the nucleotide sequence of the DNA within a gene. If the gene codes for a protein, this will affect the amino acid sequence of the protein it codes for and may change the structure and the function of the protein. These mutations are passed on to successive generations of offspring, increasing the number of individuals in a population who carry this mutation.

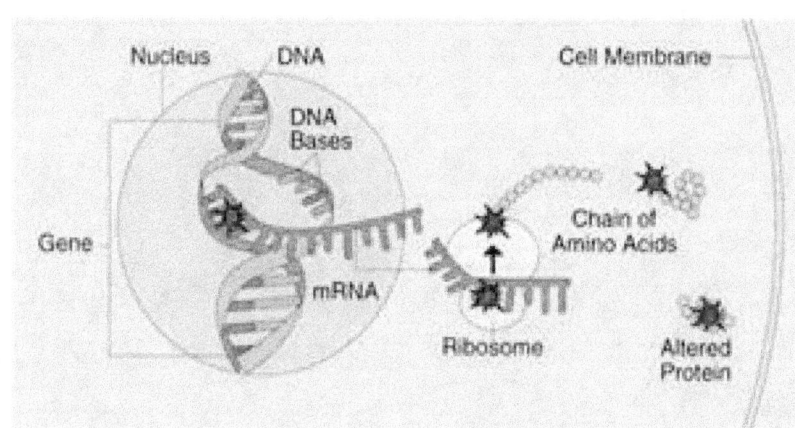

Knowledge and Comprehension:

Adaptation:

Evolve:

Natural Selection:

Environment:

Physical Adaptation:

Behavioral Adaptation:

1. What is an adaptation?

2. What is the process of natural selection.

Application, Analysis, Evaluation, and Synthesis

3. How does natural selection ensure the survivability of an organism?

4. What two types of adaptations can occur? How are they different?

5. Explain how the environment can cause adaptations to be passed down from generation to generation.

6. Why are adaptations important to organisms? Support your response by using evidence from the text.

The Transition from the Jungle to the Savanna

The first primates appeared roughly 85,000 years ago during the Cretaceous period. **Primates** evolved from mammals that lived among trees of the tropical jungles and subtropical rainforests of Northern Africa. They split, millions of years ago, into two important types of primates: prosimians and simians. The **prosimians** were primarily tree dwellers. Their food sources, small animals and insects, were located within the trees. A drastic change in the climate occurred causing a world-wide cooling trend.  This cooling trend caused the average global temperature to drop by 10 to 20 degrees Fahrenheit, allowing the onset of an ice age. As the climate changed, the trees in the jungle biome slowly disappeared. These open areas turned into savannas and grasslands.

The ancestors of the great apes and the hominids, the early **simians**, adjusted to their new environment in the

savannas by developing adaptations that help organisms to survive in a new environment. These adaptations, becoming bipedal walking on two feet, and use of tools caused the brains of these primates to reorganize and to grow in size. These early changes in locomotion freed the hands and allowed them to concentrate on making and using tools.

Creating and using tools mentally challenged early primates. This was a complex mental process that required the early primates to understand the purpose for making a tool, envision the finished product they wanted to create, and figure out different techniques of making the tool. Through the use of the tool, they could evaluate it and modify it to improve its performance.

Knowledge and Comprehension
Words to Know:

Primates:

Prosimians:

Simians:

Adaptation:

Bipedal:

1. When did the ancestors of the hominids transition from the jungle to the savanna.

2. Explain why the ancestors of the hominids were forced to move into the savannas.

3. What major adaptation did the ancestors of the hominids make as they adjusted to living in the savannas?

Application, Analysis, Evaluation and Synthesis

4. How did bipedalism, walking on 2 feet, help the hominids develop the use of tools.

5. Explain how new adaptations the hominids made initiated the reorganization of the brain and the growth of the brain.

6. Create a short storybook about how the ancestors of the hominids transitioned from the jungles into the savannas. Include pictures along with your writing.

The Transitional Species: Homo Habilis and Homo Rudolfensis

The hominids **Homo Habilis** and **Homo Rudolfensis** have been looked at as transitional species between Australopithecus Afarensis, the first known hominid, and Homo Erectus, the direct hominid ancestor to humans. Homo Habilis lived 2.33 to 1.44 million years ago (mya) and Homo Rudolfensis lived 1.9 million years ago. These two species of hominids are called **transitional species** because they have characteristics or traits in between those of early primates and the more human-like hominids that came later in the history of humans. While they were more ape-like, they also displayed traits that became more human-like as time progressed.

The adaptations that early primates and hominids such as Australopithecus Afarensis made as a result of adjusting to a new life in the savannas, continued in

Homo Habilis and Homo Rudolfensis. The two main adaptations, becoming bi-pedal, walking on two feet, and the use of tools, were important to their survival in the savannas. Their body structures such as their legs, arms, feet, and hands progressively changed as they learned how to move in different ways than their primate ancestors in the jungles.

Bipedalism, walking on feet, allowed these hominids to use their hands to make tools and to hunt with these tools. A **tool** is any implement that is used as a means of accomplishing a task. As their tools became more complex throughout time, so did their techniques for making and using these tools. Evidence of their tools, known as the **Oldowan Industrial complex** were found by Homo Habilis fossils by scientists, in Lake Turkana, Kenya and the Olduvai Gorge in Tanzania. Some of the tools they created including choppers, scrapers, and pounders made of stone. They were used for hunting small game animals, processing and cutting raw meat, as well as, for making tools. The oldest

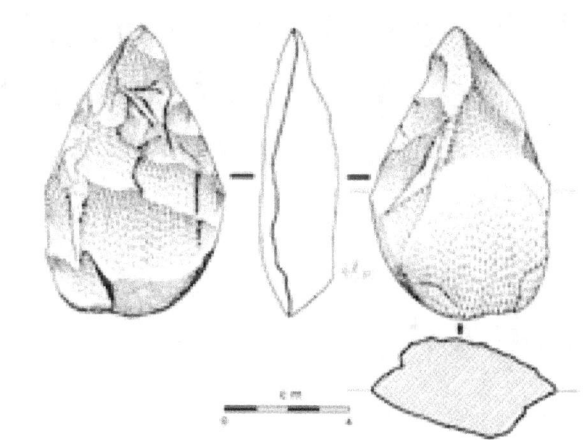

Oldowan tools were found in Gona, Ethiopia at have been dated back to 2.6 million years old.

They were able to change and modify their tools based on their performance. Improvements in language and communication between members of the hominid groups allowed information about how to make tools and use tools, hunting strategies, and where to find resources to be shared. This  adaptation also helped the hominids to improve their diets as well. They foraged for meat from wild game, as well as, plants. In response to a more nutritious diet, the use of their brains for making and using tools, and the development of language, their brains continually reorganized and grew exponentially.

Source:

http://www.sabali.co/research/hominid-brain-development.php#introduction

Knowledge and Comprehension
Words to Know:

Homo Habilis:

Homo Rudolfensis:

Hominids:

Transitional species:

Bipedalism:

Tool:

Oldowan Industrial Complex:

1. Which hominids were considered transitional species?

2. What is a tool?

Application, Analysis, Evaluation and Synthesis

3. Explain why Homo Habilis and Homo Rudolfensis are considered transitional species?

4. Describe the tools of Homo Habilis. Where in Africa have they been found? What were they used for?

5. How did the development of language affect tool making and hunting within groups of Homo Habilis hominids?

6. Which adaptations allowed the Homo Habilis brain to continue to develop and grow?

Homo Erectus

Homo Erectus, also known as Homo Ergaster, is a hominid species that lived about 1.8 - 0.1 million years ago or 100,000 years ago. The species had a cranial capacity and brain size larger than Homo Habilis. The size of their brains ranged from 850 to 1100 cubic centimeters in volume. It is believed that Homo Erectus was the ancestor of Homo heidelbergensis, Homo Neanderthalensis, and Homo Sapiens.

Many Homo Erectus **fossils**, or bones found within the rock layers, have been found in Africa such as Lake Turkana, Kenya. Fossils have been also been found in the Republic of Georgia, Java in Indonesia, Beijing in China, Vietnam, India, and Sri Lanka. This evidence has lead to the hypothesis that this hominid originated in Africa and then migrated into Europe, Asia, Indonesia, and India.

The Sahara Pump theory is a possible explanation for the migration of the Homo Erectus Species between Africa, Europe and Asia through a land bridge located in the **Levant** or the region between the Mediterranean Sea and

Iraq. This migration was initiated and caused by the transition of Northern Africa and Arabia from **pluvial periods** wet and rainy conditions lasting for thousands of years to desert conditions that also lasted for thousands of years. The wet and arid conditions followed a 26,000 year cycle. From 1.8 to .8 million years ago, geologic uplift of the Nile River region known as the **Nubian Swell**, caused the water flow of the Nile to be slowed down significantly. The animal and plant life, as a result, changed drastically during the times of desert conditions. The savannas, grasslands and large lakes and rivers slowly disappeared. Hominids, such as Homo Erectus, **migrated** or moved to central and southern Africa, as well as, migrating out of Africa into the Levantine land bridge to survive. From this region, hominids migrated into Europe, and into Asia.

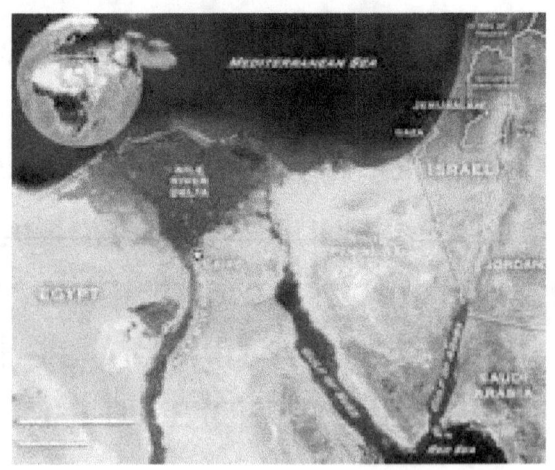

Homo Erectus was known to use tools such as handaxes to cut meat. They also gathered nuts, berries, and fruits as food. They may have been the first hominids to live in

hunter-gatherer band-societies. They are also thought to be the first to hunt in coordinated groups and use complex tools. These tools have also been found along with their fossils.

They have demonstrated that they were able to control fire. Evidence or **artifacts**, or hand-made implements, at various fossil sites for this in East Africa was found in the form of red clay sherds that been heated in "hearth-like depressions. Clay pottery, in ancient times, was hardened by heating it up to 700 degrees Fahrenheit. Burned bones and plant ash has been found at other fossil sites in Israel and dated from 790,000 to 690,000 BP.

Source:

http://en.wikipedia.org/wiki/Homo_Erectus

Knowledge and Comprehension
Words to Know:

Homo Erectus:

Fossils:

Levant:

Pluvial Periods:

Nubian Swells:

Migrate:

Hunter gather Band-Societies:

1. When did Homo Erectus live?

2. Where were the fossils of Homo Erectus found?

Application, Analysis, Evaluation and Synthesis

3. Describe the lifestyle of Homo Erectus.

4. Explain why Homo Erectus migrated out of Africa.

5. What artifacts did scientists find that suggests that they knew how to control fire?

Homo Heidelbergensis

Homo Heidelbergensis, also known as Homo Rhodesiensis, is a hominid species that lived about 1.3 - 200,000 years ago. Fossils in South Africa indicate they lived between 500,000 to 300,000 years ago. Fossils of this hominid have also been found in Europe, Asia, and Africa. By comparison, many Homo Erectus **fossils** have been found in Africa such as Lake Turkana, Kenya, and in the Republic of Georgia, Java in Indonesia, Beijing in China, Vietnam, India, and Sri Lanka. It is theorized that Homo Heidelbergensis is a descendant of **Homo Ergaster** or Homo Erectus of Africa . The first fossil, or preserved bone found within rock layers, of this species was found near Heidelberg, Germany. It is also believed by scientists that Homo heidelbergensis is the direct ancestor of Homo Neanderthalensis, the Denisovans in Eastern Asia, and the Homo Sapiens.

Members of Homo Heidelbergensis had a brain size that was 1100 - 1400 cubic centimeters in volume, matching that of Homo Sapiens with a brain size of 1,350 cubic centimeters. The males had a height of about 5 feet 9 inches, and the females, a height of 5 feet 2 inches.

It has shown that the Neanderthals shared many of the physical features of Homo Heidelbergensis with a few differences. **Neanderthals** are shorter in stature, stockier, had large facial brow ridges, lack of a dominant chin, and a face that was slightly protruding.

It is theorized that the Neanderthals diverged from one population of Homo Heidelbergensis while they were in Europe about 300,000 years ago. Cro-Magnon man Homo sapiens, on the other hand, diverged from another population of Homo heidelbergensis in Africa, before they migrated out of Africa into Europe about 200,000 years ago. Hominid fossils found in Atapuerca, Spain and Kabwe, Zambia in South Africa represent members from two branches of the Homo Heidelbergensis tree. The fossil found in Kabwe, known as **Rhodesian man** displays features that are a mixture of Neanderthal and Homo Sapien. It has a broad face, a large nose, and thick brow ridges like the Neanderthals. Its body has features in-between those of Neanderthal and Homo Sapiens. It is theorized by scientists that this hominid has evidence of DNA **admixture** or mixing resulting from interbreeding of Neanderthals and Homo Sapiens before each migrated

out of Africa. This fossil has been nicknamed "African Neanderthal" because of its traits.

The divergence of Neaderthals and Homo Sapiens from their common ancestor, Homo heidelbergensis can be interpreted as an example of **divergent evolution**. The traits and DNA between the two hominids were in the process of evolving and becoming different from each other through time. Because they were able to interbreed with each other and leave viable offspring before Neanderthals went extinct, the two hominids may have been members of the same species. Their DNA is 99.87% similar. The similarity between the DNA of modern humans is up to 99.5%. Each hominid may have been on the road to becoming distinct species through divergent evolution.

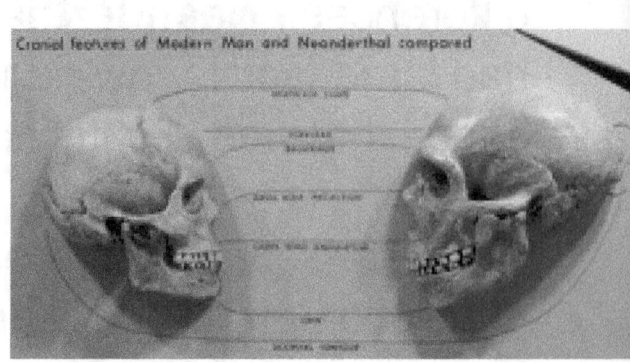

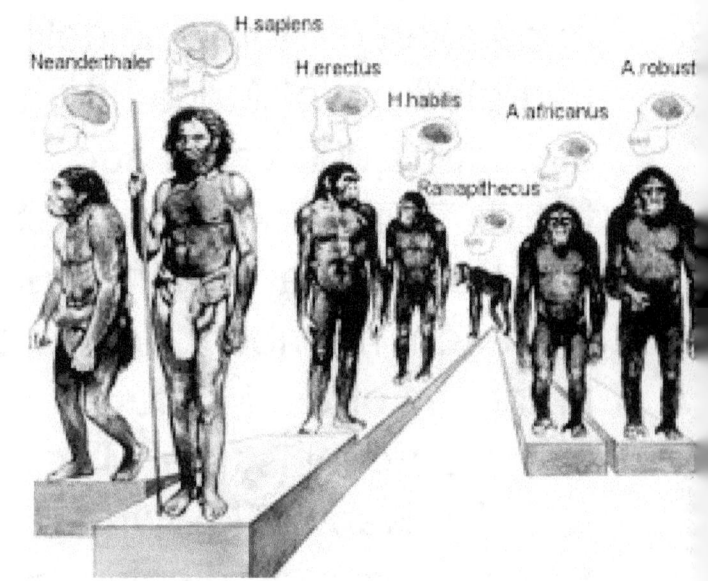

Source:

http://en.wikipedia.org/wiki/Homo_heidelbergensis

Knowledge and Comprehension
Words to Know:

Homo Heidelbergensis:

Fossils:

Homo Ergaster:

Neanderthals:

Rhodesian Man:

Admixture:

Divergent Evolution:

1. Describe who Homo Heidelbergensis was?

2. Who is the ancestor of Homo Heidelbergensis?

3. How many populations of Homo Heidelbergensis have been found in Africa and Asia? Who are their descendants?

4. How can admixture or the mixing of DNA be accomplished between two hominids?

5. Do you agree or disagree that Rhodesian Man is the offspring of Neanderthals and Homo Sapiens?

6. Demonstrate how divergent evolution plays a role in the development of Neanderthals and Homo Sapiens.

Homo Neanderthalensis

Homo Neanderthalensis is a hominid species that lived from 500,000 to 30,000 BC. Now extinct, these hominids are regarded by some scientists as a sub-species of modern humans. A **sub-species** is a taxonomic rank comes under species classification and represents different groups of the same species that live in different geographical locations or may represent wild and domesticated varieties of the same species of organism. Recent mitochondrial DNA evidence has shown that they interbred with Homo Sapiens and another hominid known as the Denisovans. This evidence may suggest that Homo Neanderthalensis, Homo Sapiens, and the Denisovans may be sub-species of modern humans especially if they left viable and fertile offspring as stated in the definition of "**species**.". In this case, they are regarded as members of the same species.

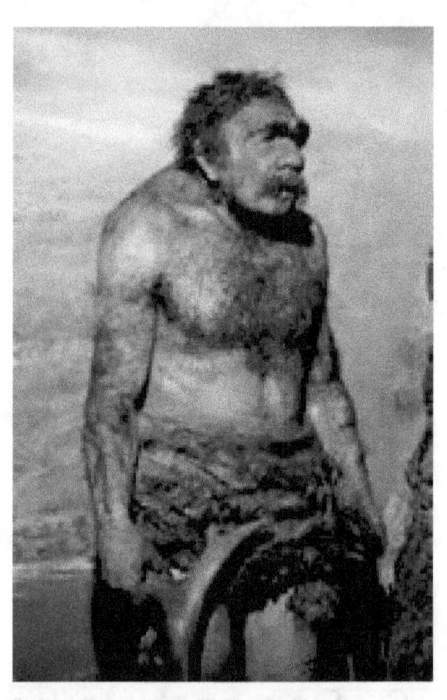
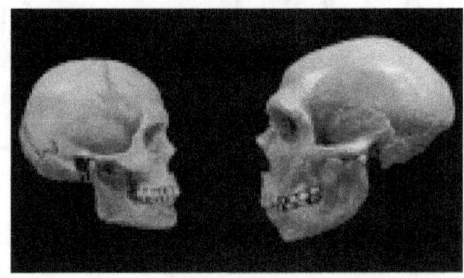

Homo Neanderthal fossils have been found in western Europe and Asia. They have been named after a fossil site in the Neander Valley in Germany. The first proto-Neanderthals are believed to have lived in Eurasia from 600,000 to 350,000 BC. A Cro-magnon man fossil found in a rock shelter in Lagar Velho, Portugal, dates back to 24,000 BC and has some Neanderthal traits. **Cro-magnon man** was an early modern human or Homo Sapien. This fossil shows characteristics between the Cro-magnon man and the Neanderthals suggesting that they interbred with each other. Today Homo sapiens carry 1-4% Neanderthal DNA worldwide.

The Neanderthals had an average brain capacity of 1600 cubic centimeters. Their brain size was clearly larger than modern humans today. Males had an average height of 5 feet 5 inches and females had an average height of 5 feet. Their rib cage was larger than modern humans and were relatively stronger.

They have also been shown to make and use a group of tools known as **Mousterian tools**. These tools were primarily made of flint and were found in the Le Moustier rock shelter site in the Dordogne area of France. **Flint tools** are stone tools that that are made from flaking and chipping off pieces using a hard hammerstone and pressure. They used specific tools such as hand axes, arrowhead points, and scrapers.

Sources:

http://en.wikipedia.org/wiki/Neanderthal

http://ngm.nationalgeographic.com/2008/10/neanderthals/hall-text

Knowledge and Comprehension:

Homo Neanderthalensis:

Sub-species:

Species:

Cromagnon Man:

Mousterian Tools:

Flint Tools:

1. Describe who the Neanderthals were.

   Describe where they lived and where their fossils have been found.

   Describe their tools.

2. Explain what a sub-species is.

Application, Analysis, Evaluation, and Synthesis

3. How are Neanderthals different from modern humans?

4. Use your imagination and describe what Neanderthals would have been like if they had not gone extinct.

5. Write an argument to explain why you agree or disagree with the statement: Neanderthals and modern humans are sub-species. Use evidence from the text to support your answer.

Differences in Adaptations among Neanderthals and Homo Sapiens

Adaptations are traits and characteristics that allow organisms a survival advantage with a particular environment. The adaptations that are developed is an organism's way of responding to changes within the physical environment, as well as, within the internal environment. The body of an organism, in an effort to maintain **homeostasis** or a stable internal environment, regulates many biochemical reactions that cause important functions within the cells to occur properly. It has been proven that changes in the climate such as light, temperature and humidity cause changes in the traits of organisms. It has been a dramatic catalyst for change within the genomes of the hominids and Homo Sapiens.

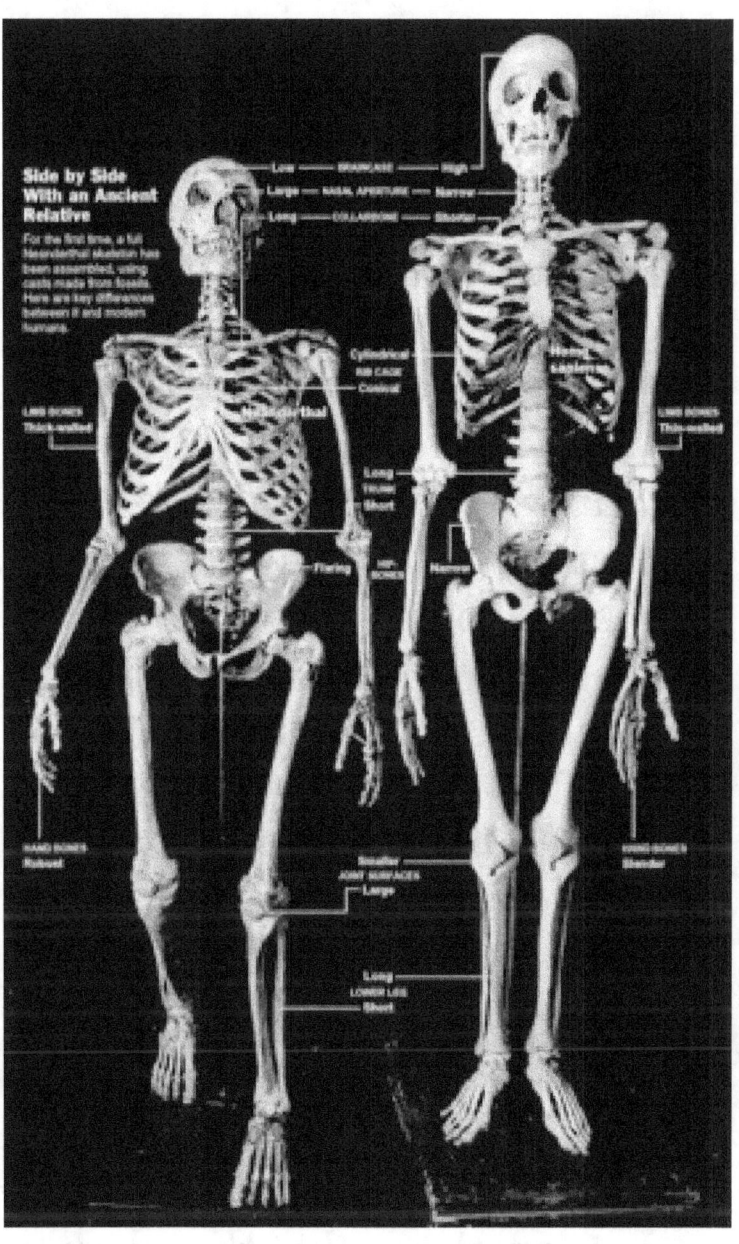

Neanderthals and Homo Sapiens, even though they were 99.84% similar to one another, had morphological and structural differences in anatomy. These two hominids adapted to different climates and regions of the world over time. The Neanderthals moved North into Europe and Asia while modern humans stayed in Africa until about 50,000 years ago. The Neanderthals that moved North out of Africa left a warm, subtropical region for a new environment that would drastically be altered by an ice age. Ice sheets, at one time, reached into the southern limits of the Netherlands, Germany and Poland on the continent of Europe. Many **Neanderthal** groups lived within this region. Adaptations that they developed include lighter skin color due to the reduced amount of light at these latitudes. They also developed more hair and thicker hair due to the drop in the average temperature of the atmosphere. They were also shorter, stockier body shape, and shorter, stronger

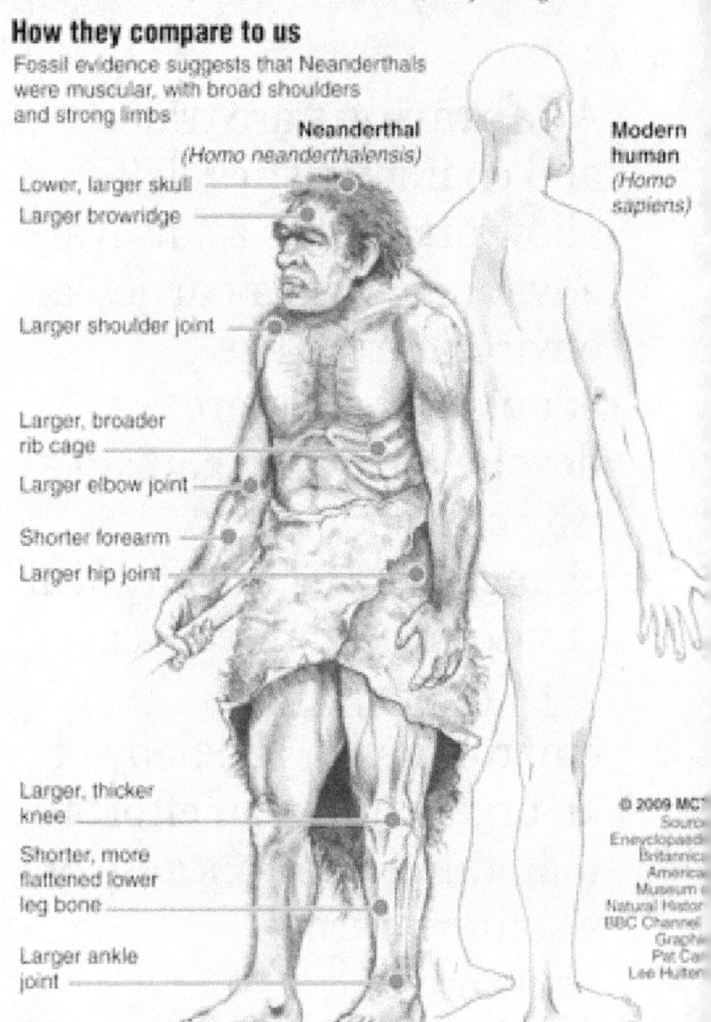

arms and limbs. Their bodies were adapted to a physically demanding environment that included the hunting of large game animals requiring strength to lift and move heavy objects from place to place.

Homo Sapiens, during this time, endured many alternations in wet, rainy climate and hot, dry desert-like conditions. **Climate** refers to the long term weather patterns of a region. Homo sapiens in Africa developed darker skin color due to the increase of sunlight near the equator of the Earth. They also had less hair and taller, leaner bodies due to the warmer temperatures. They changed their hunting patterns to include big game and small game animals as well. Their bodies were adapted to running fast after prey over long distances.

Cro-Magnon man as shown in Richard Leakey's book "Origins" published in 1977.

Researchers have discovered that the differences in the anatomies of the Neanderthals and Homo Sapiens can be explained through gene expression. It has been shown that there are 2,200 similar genes between these two hominids. Some of these genes are switched on in Neanderthals and switched off in modern humans and vice versa. The switching on and off of these genes depended on where they lived and the climate they were exposed to in these locations.

Sources:

http://www.huffingtonpost.com/2014/04/18/dna-neanderthals-modern-humans-genes_n_5168730.html
http://www2.lbl.gov/Science-Articles/Archive/Genomics-Neanderthal.html

Knowledge and Comprehension:

Adaptations:

Homeostasis:

Neanderthal:

Homosapiens:

Climate:

1. What are adaptations?

2. Explain why adaptations occur.

3. What can affect homeostasis within an organism?

Application, Analysis, Evaluation, and Synthesis

4. Explain how changes in the climate affected the development of adaptations that occurred within Homo Sapiens and Neanderthals?

5. If Homo Sapiens and Neanderthals are 99.84% similar in DNA show how they are different in structures.

The Denisovans

The Denisovans are an extinct hominid species that ranged from southeast Asia to Siberia. The first Denisovan fossil found in a cave in the Altai Mountains in northeastern Asia. The **fossil** remains were found alongside Neanderthal and Homo Sapien fossils and dates back to 41,000 years ago. Human artifacts found in this cave date back to 125,000 years ago.

Denisovans are shown to have a common origin with **Neanderthals** and Homo Sapiens. These hominids, including Homo Sapiens, all have a common ancestor known as **Homo Erectus**. The Denisovans diverged from the human lineage about 600,000 years ago and from the Neanderthals about 200,000 years ago. This makes the DNA of the Denisovans closer to Neanderthals than Homo

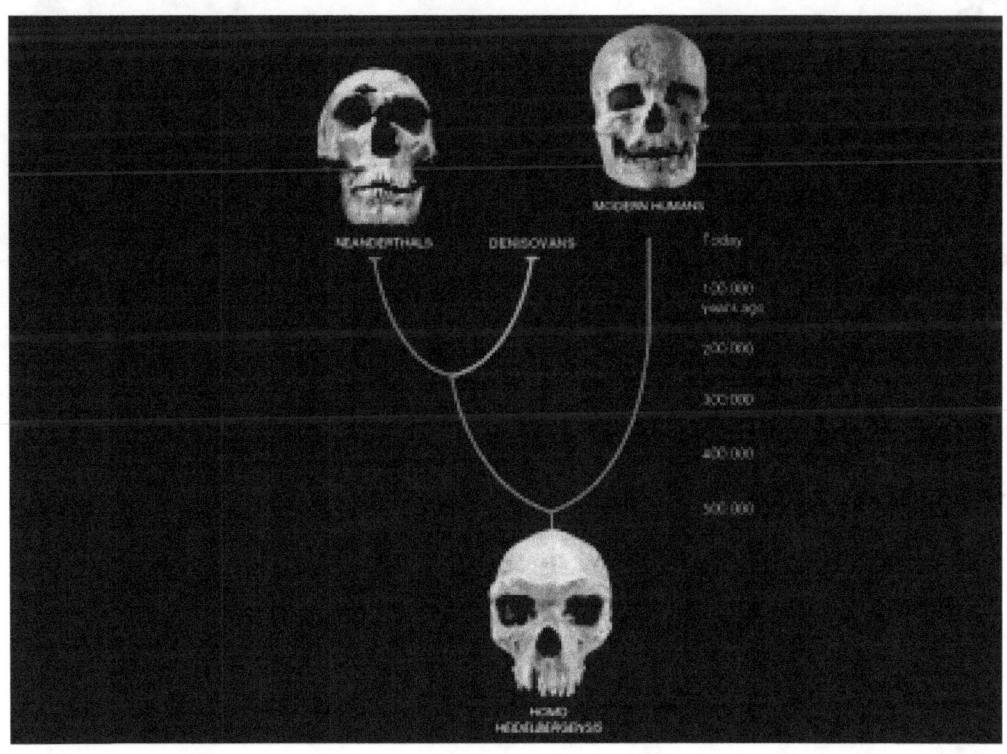

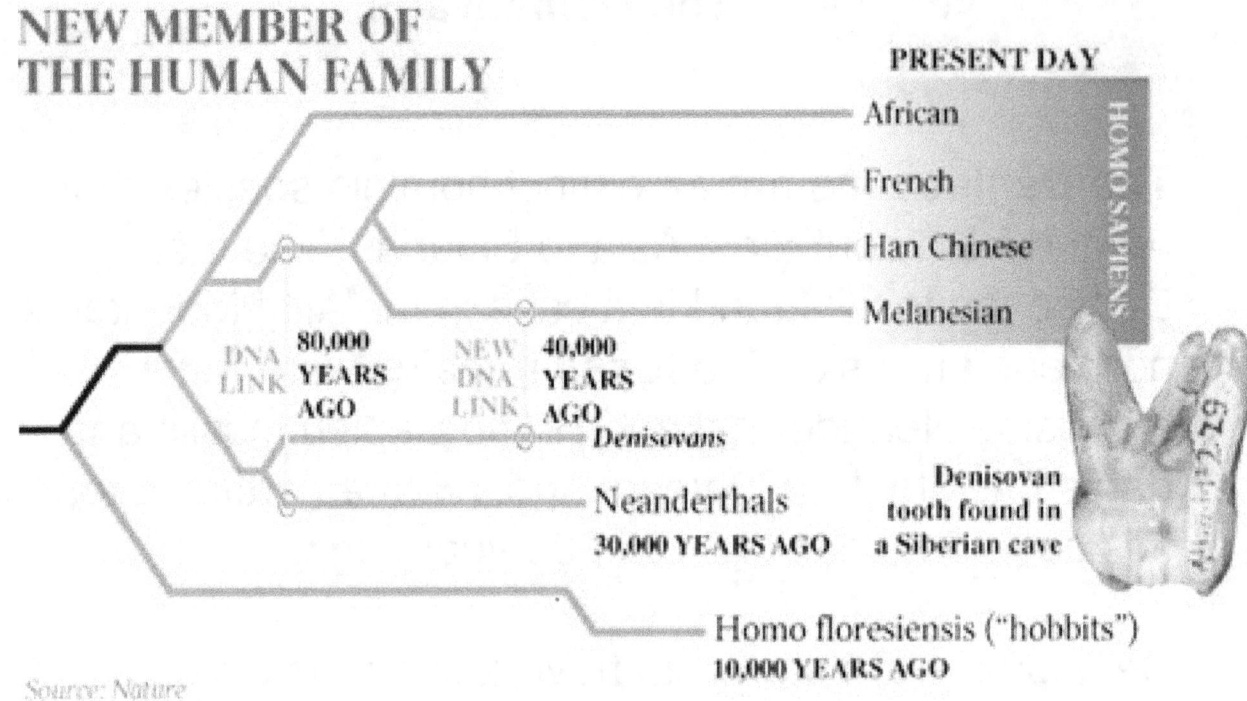

Sapiens. Because of this, they have been nicknamed by scientists as the "Asian Neanderthals." It is believed by scientists that **Homo Heidelbergensis,** Homo Erectus found in Europe, was the direct ancestor of the Neanderthals, the Denisovans of Eastern Asia, and Homo Sapiens.

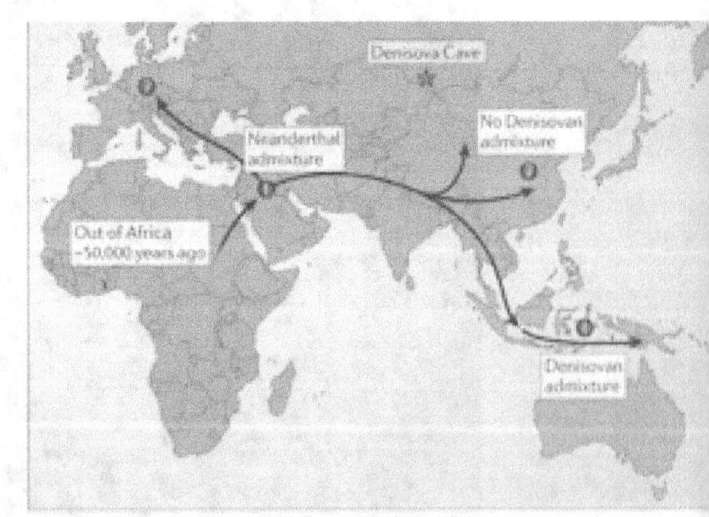

3-5% of the DNA of the Melanesians, Polynesians, Burmans, the Malays, and the Aborigines of Australia today is derived from the Denisovans. This evidence suggests admixture of Homo Sapien DNA with Denisovan DNA. **Admixture** is the attaining of DNA from a different source through interbreeding. 17% of their DNA, from specimens in the Altai cave, derived from Neanderthals. This evidence suggests that the genome of Homo Sapiens in Asia, Australia and Indonesia is a **mosaic** or a mixture of DNA from different hominids.

Source:
http://en.wikipedia.org/wiki/Denisovan

Knowledge and Comprehension
Words to Know:

Denisovans:

Fossils:

Neanderthals:

Homo Erectus:

Homo Heidelbergensis:

Admixture:

Mosaic:

1. Who are the Denisovans?

2. Where was the first Denisovan fossil found?

3. Which hominid is the ancestor of the Denisovans, the Neanderthals, and Homo Sapiens?

Application, Analysis, Evaluation and Synthesis

4. What is admixture? What proof is there today of admixture between the Denisovans and other hominids?

5. Why is the genome of Homo Sapiens is a mosaic or a mixture of DNA from different hominids? What is the evidence for this?

Homo Floresiensis

Homo Floresiensis is a hominid species that lived on the island of Flores in Indonesia. The island is east of Bali, Indonesia which is in-between Australia and Asia. Fossilized skeletons in a cave at this location date back to 18,000 years BC. It is believed by scientists that these hominids **migrated** from Asia to Indonesia and Australia 50,000 years ago. They lived from 95,000 to 13,000 years ago and existed at the same time as modern humans.

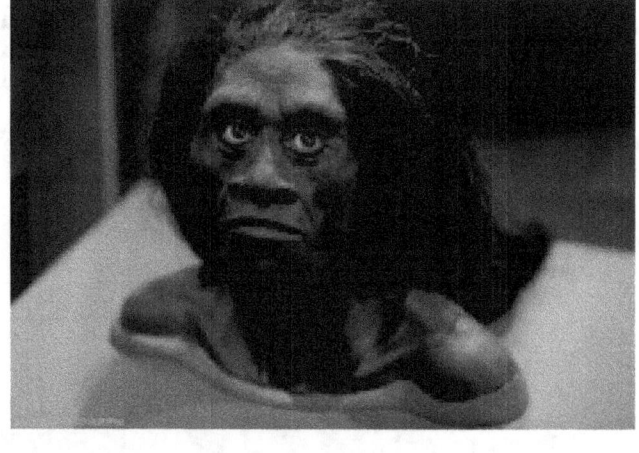

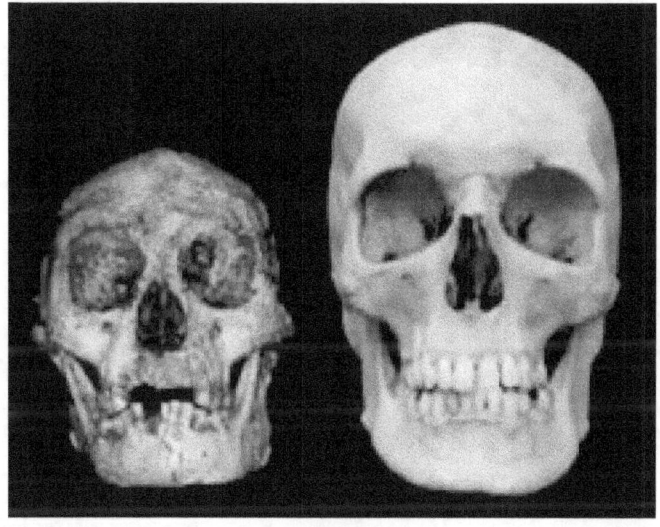

They have also been found alongside stone tools such as spear points, blades, and perforators. These primitive tools were created and used for hunting and processing foods such as meat and plants. Scientists consider these tools **artifacts**, or hand-made implements used for a specific purpose. They help to date the fossil remains of the hominids.

The individuals of this group were only 3 feet 5 inches tall and weighed about 55 pounds. They have been dubbed **"hobbits"** because of their small stature and height. Their brain sizes were as large as the size of a chimpanzee. Members of this species lived with wild animal species such as Komodo dragons, giant rodents, 10 foot lizards and pygmy elephants.

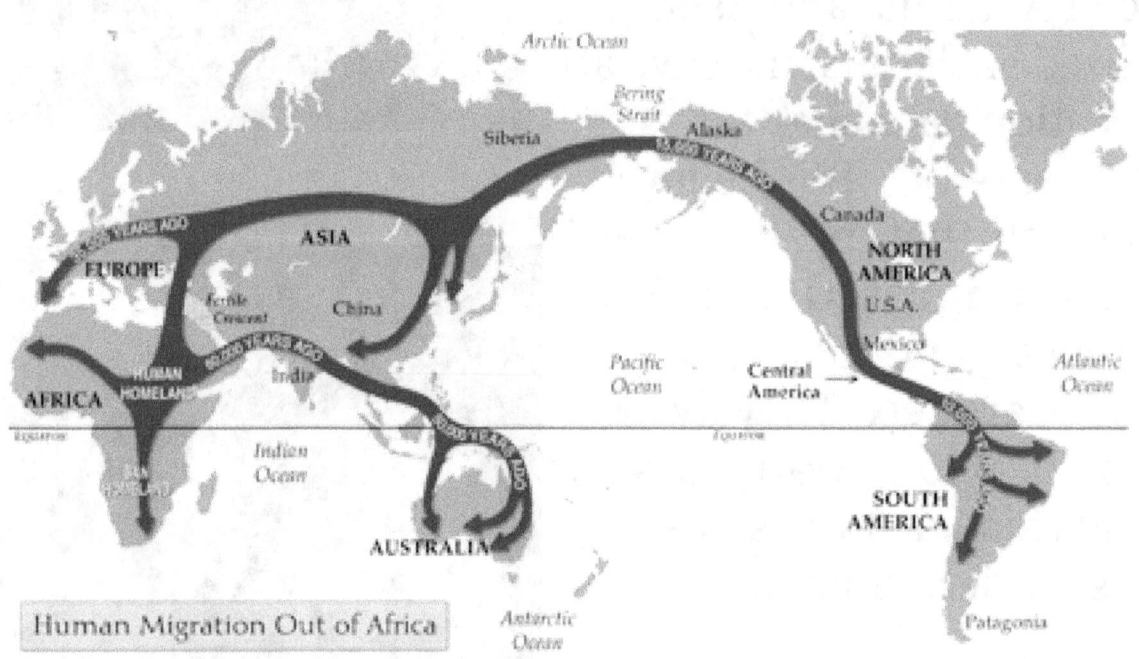

Scientists agree that the finding of Homo Floresiensis fossils on a **remote** or isolated island near Indonesia is amazing. A probable explanation for this is that 50,000 years ago the Earth was experiencing an ice age that locked up ocean water in the form of glaciers at the North and the South poles. The sea levels would have been more than 400 feet lower than they are today. It is possible that these hominids migrated on foot from the Continent of Asia, into Indonesia, the island of Flores, then Australia.

When the ice began to melt, the sea levels would have risen, cutting off the island from the Asian mainland.

Source:

http://news.nationalgeographic.com/news/2004/10/1027_041027_homo_floresiensis.html

Knowledge and Comprehension

Homo Floresiensis:

Migrate:

Artifact:

Hobbits:

Remote:

1. Write a description about Homo Floresiensis.

2. Why are these hominids called "hobbits"?

Application, Analysis, Evaluation and Synthesis

3. What types of tools did this hominid make and use?

4. What types of animals lived on the island of Flores along with this hominid? Would they have hunted these animals for food? If so, what tools would they have used? Draw these tools and write down how they could have been used them.

5. What is a possible explanation for how Homo Floresiensis became isolated on the island of Flores.

Co-existence:
Homo Sapiens and Other Hominids

Hominids have graced the planet with their presence for over one million years. We know hominids exist through the evidence they have left behind throughout time. Different types of hominid species have been found all over the world. The evidence that continues to be discovered paints a different picture and tells a different story. We are still piecing together a puzzle of human development that we are only starting to comprehend.

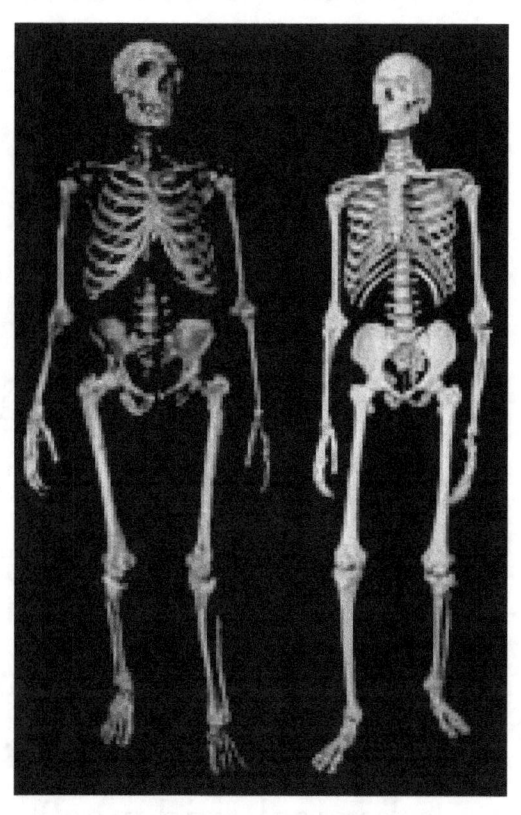

What is intriguing is to analyze the myths alongside the popular scientific theories that suggest how the human race came into being. The most plausible explanation for the development of the human species lies with Charles Darwin. As explained by his world renowned study, the *Origin of Species (1859)*, he theorizes that humans are a product of evolution. Natural selection played a big part in the differentiation of the animal kingdom to eventually, through millions of years, give rise to its crowning achievement…humans. Us. How is this possible?

The origin of mankind is not treated as a singular, special event of creation as stated in many of the creation myths and stories found all over the world. The account of creation, as stated in the Bible, tells us that the Gods created humans in their image. This miraculous event took place in one day. Creation myths from other cultures recount almost the same idea, but in different ways. This idea is that mankind was created by the Gods. If this is actually the truth of our creation, where is the physical evidence?

Charles Darwin was a naturalist and based his theory for the "creation of mankind" from evidence in the fossil record. What he was able to decipher was the comparison of characteristics among different animal species and establishing a biological relationship between species. What he was able to conclude was that species of animals that shared more characteristics between one another were more closely related to one another. This conclusion was based out of the studies of different bird and animal species he studied on his voyages to the Galapagos Islands in South America.

The first human fossil find in 1857 proved to be a challenge for naturalists. The remains of a skull capwas found at the Feldhofer Grotto within the Neander Valley by two miners in Germany. The skull looked human but contained a few features that were different than modern humans. The bone of the skull was thick, and it contained a thick ridge where the eyebrows are located.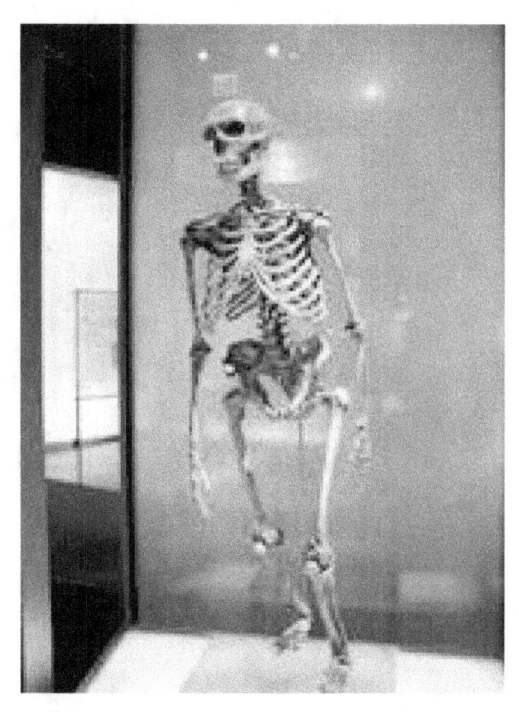
Herman Schaaffhausen, one of the German naturalists, was convinced that the skull belonged to an extreme member of the species Homo Sapiens. He stated that it was a "recent barbarian," maybe a member of one of the wild tribes living in the region. He ignored the fact that the skull was found in a cave with two extinct animal species; the cave bear and the mammoth.

Upon this incredible find, Darwin considered the human skull and began to contemplate the origins of humans. He published a book called The Descent of Man and Selection in Relation to Sex" in 1871 and stated that the known evidence is consistent with humans having evolved from a

common ancestor with apes. He also theorized that their place of origin was Africa and humans have evolved into their present form since that time. Alfred Russel Wallace, his correspondent , did not agree. He was convinced that humans were the work of divine intervention. He thought the human brain was far more special, powerful, and more advanced than that of an ape.

In 1886, Neaderthal fossils were found in Spy, Belgium. A jaw and a partial skeleton were dug out from ancient sediment. Eugene Dubois traveled to Indonesia the year after and found a human fossil in eastern Java, Indonesia near the Solo River four years later. He had found the remains of *Pithecanthropus erectus* or "upright ape-man." A being that was to small to be human and was neither fully apelike. This was the first member of the Homo Erectus species that was found anywhere

in the world.

At present, 20 different hominid species have been found worldwide. The oldest hominid dates back 6 million years. They suggest an African origin as suggested by Darwin. The fossils also show evidence of a non-linear hominid development with many different species of hominids existing at the same time, except for the last 30,000 years. From 400,000 to 30,000 years ago, the Neanderthals, thick boned, hunter-gatherers living in Europe and Asia, were the last human group to become extinct. Svante Pääbo, from the Max Planck Institute for Evolutionary Anthropology in Leipzig, Germany, states that the Neanderthals split from modern humans less than 500,000 years ago. Their common ancestor lived in Africa giving rise to the modern human.

50,000 years ago, The Neanderthals and the Denisovians, were found living alongside the Denisovians and modern humans in caves in Denisova, Asia during the last ice age. Denisova is located in the Atlai mountain range in east-central Asia. The Altai mountains is in a region where China, Mongolia, and Russia come together. The Denisovians, an archaic group of humans, have a different DNA profile than modern humans. Neanderthals and the Denisovians were able to interbreed and have viable offspring.

Homo Sapiens and the Neanderthals were also able to interbreed as well. Montgomery Slatkin and Anna-Sapfo Malaspinas, theoretical geneticists from the University of California, Berkeley, found that this interbreeding between Neanderthals and Humans occurred about 65,000 - 90,000 years ago. They are thought to be two members of the same species or sub-species. There is evidence that Homo Sapiens and Neanderthals genetically mixed, reproduced, and produced viable offspring. Only members of the same species can reproduce and have viable offspring. 2 to 4% of human DNA contains Neanderthal DNA. Most of this ancient DNA codes for elements within the immune system.

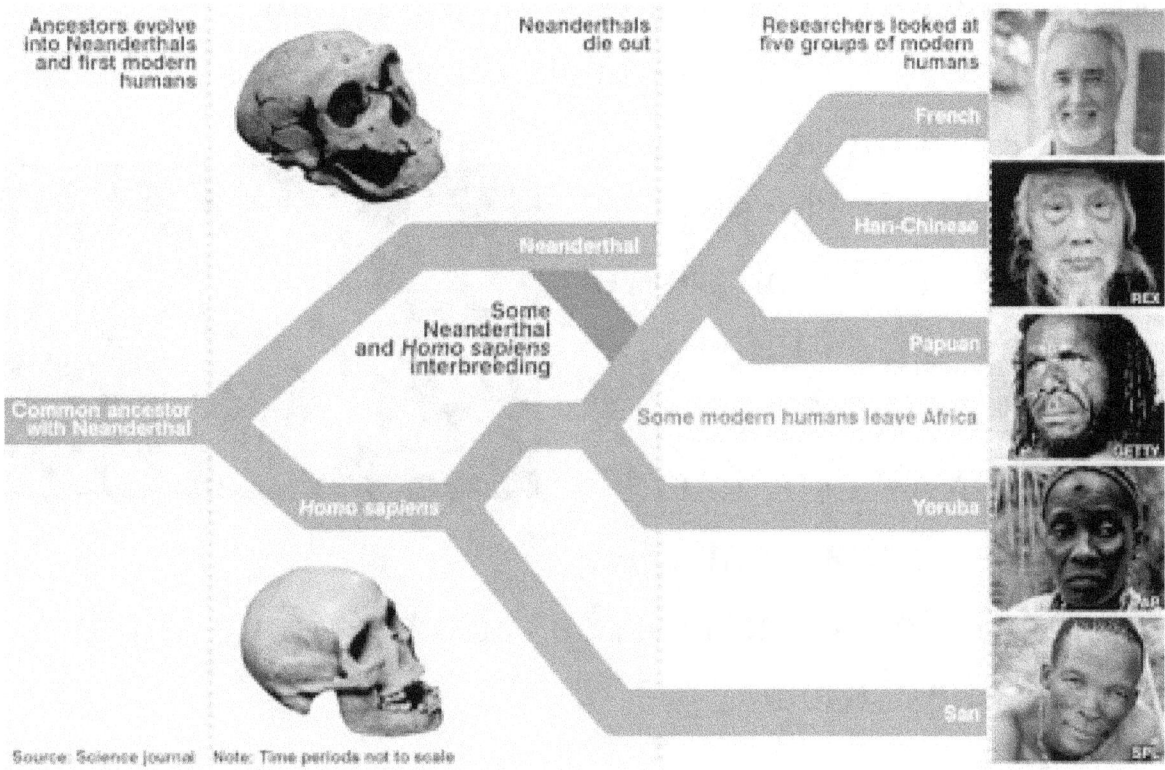

On March 28, 2013 it was announced that the fossil skeletal remains of a human-Neanderthal hybrid dating to 30,000 - 40,000 years old was found in a rock shelter named Riparo de Mezzena, in Northern Italy. The research team discovered the find in an area of Italy where Neanderthals and Humans once co-inhabited. Scientists have also found fossil remains of bot humans and Neanderthals in several rock shelters in France.

The Denisovans, another potential subspecies of humans also interbred bred with Homo Sapiens according to according to

Pääbo and Reich. Fossils of Homo Sapiens and Denisovans were found together, in cave, at a fossil site in the Atlai mountain range in east-central Asia. The Altai mountains is in a region where China, Mongolia, and Russia come together. 4 - 6% of Denisovan DNA can be found in the people who presently live in Melanesia. Denisovan DNA has also been found in the Aborigines of Australia.

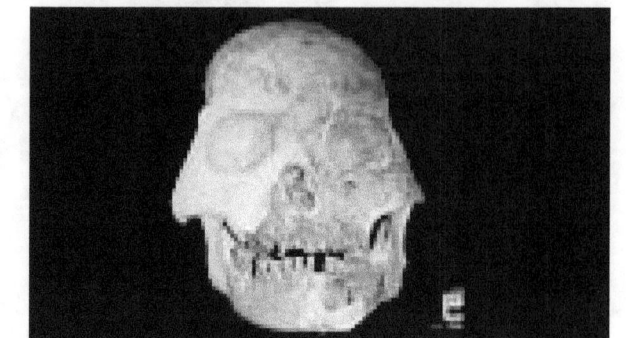

Sources:

http://evolution.berkeley.edu/evolibrary/article/history_17
http://www.sott.net/article/233250-Ancient-DNA-Reveals-Secrets-of-Human-History
http://news.msn.com/science-technology/skeletal-remains-may-point-to-human-neanderthal-interbreeding

Write an Argument

Do you agree or disagree with the claim "Humans are a product of different hominids." Find evidence in the text to write an argument that supports your response.